Biogas - Basics, Integrated Approaches, and Case Studies

Edited by Abd El-Fatah Abomohra and El-Sayed Salama

Published in London, United Kingdom

IntechOpen

Supporting open minds since 2005

Biogas - Basics, Integrated Approaches, and Case Studies
http://dx.doi.org/10.5772/intechopen.95674
Edited by Abd El-Fatah Abomohra and El-Sayed Salama

Contributors
Donald Kukwa, Maggie Manimagalay Chetty, Zikhona Tshemese, Ndumiso Duma, Denzil Estrice, Adrian Eugen Cioabla, Francisc Popescu, Sharda Dhadse, Shanta Satyanarayan, Sammy N. Aso, Simeon C. Achinewhu, Madu O. Iwe, Venko Beschkov, Jean Pierre Doussoulin, Cristina Salazar Molina, Abd El-Fatah Abomohra, El-Sayed Salama

First published in London, United Kingdom, 2022 by IntechOpen
IntechOpen is the global imprint of INTECHOPEN LIMITED, registered in England and Wales, registration number: 11086078, 5 Princes Gate Court, London, SW7 2QJ, United Kingdom
Printed in Croatia

British Library Cataloguing-in-Publication Data
A catalogue record for this book is available from the British Library

Additional hard and PDF copies can be obtained from orders@intechopen.com

Biogas - Basics, Integrated Approaches, and Case Studies
Edited by Abd El-Fatah Abomohra and El-Sayed Salama
p. cm.
Print ISBN 978-1-80355-108-1
Online ISBN 978-1-80355-109-8
eBook (PDF) ISBN 978-1-80355-110-4

For EU product safety concerns:
IN TECH d.o.o., Prolaz Marije Krucifikse Kozulić 3, 51000 Rijeka, Croatia, info@intechopen.com or visit our website at intechopen.com.

Meet the editor

Abd El-Fatah Abomohra is a professor of Environmental Engineering and the director of the New Energy and Environmental Laboratory (NEEL), Chengdu University, China. He is a DAAD fellow and performed postdoctoral research at Hamburg University, Germany, and Harbin Institute of Technology, China. He was named in the world's top 2% of scientists for two consecutive years. His team is primarily working on green energy production from different biomass feedstocks.

El-Sayed Salama is a professor in the Public Health School, Lanzhou University, China. He is also a director of the Green Environmental & Energy Laboratory (GEEL) and a distinguished expert of science and technology in the ecological industry, Gansu Province. His GEEL research work focuses on bioenvironmental science and bioenergy. He has several journal publications to his credit.

Contents

Preface

The exponential growth of the global population and concurrent fast industrialization has led to the massive generation of municipal wastes, raising challenges of safe disposal. The proper management of municipal wastes through recycling is an essential approach for global sustainable development. So far, many countries have established regulatory guidelines for different waste management routes and pollution control measures. However, most of the applied routes are waste dumping, composting, or direct discharge into water bodies without adequate pretreatment, which seriously threatens the environment and humans. Thus, proper waste segregation and separation provide an efficient option for waste conversion into energy. On the other hand, energy demand correlates with population growth. Thus, global energy demand and environmental pollution are two inevitable issues that dictate the need to find alternative energy sources. Waste-to-energy is a widely used process for efficient waste management that is attracting much attention. For almost two decades, biofuel production from biowastes has been of paramount importance. In general, it is widely accepted that biowaste-derived fuels can reduce the current dependence on fossil-based products. Among different biofuel production routes, anaerobic digestion is, by far, the single most important technology for providing clean renewable biogas to millions of people in the rural areas of developing countries. Anaerobic digestion technology has several inherent benefits ranging from generating renewable energy, remediating biowaste and curtailing CO2/CH4 emissions to improving health/hygiene and overall socio economic status of rural communities in developing nations.

This book is an extension of our previously published book entitled *Biogas - Recent Advances and Integrated Approaches*. It provides new integrated approaches and case studies on biogas production. The book is divided into two main sections. The first section discusses the basics of biogas production from different feedstocks and the role of the microbial community, with the possible utilization of anaerobic digestate as a biofertilizer. The second section includes case studies and discusses the economic feasibility of biogas production from municipal waste.

We would like to express our gratitude to all the contributing authors. We also wish to thank the author service managers at IntechOpen for being generously helpful throughout the publication.

Abd El-Fatah Abomohra
Chengdu University,
Chengdu, China

El-Sayed Salama
Lanzhou University,
Lanzhou, China

Section 1

Basics of Biogas Production

Chapter 1

Introductory Chapter: From Biogas Lab-Scale towards Industrialization

El-Sayed Salama and Abd El-Fatah Abomohra

1. Introduction

Production and consumption of food and the exploitation of fossils in the past several decades due to globalization resulted in the depletion of fossil fuels and severe environmental pollution [1, 2]. The emission of greenhouse gases (GHGs) in such an increasing trend causes global warming that devastates aquatic and terrestrial ecosystems. Around 88% of energy worldwide is provided by fossil fuels despite their damage to the environment [3, 4]. In 2019, oil production reached 4484.5 Mt, with natural gas reaching 3989.3 billion cubic meters [5]. Total global coal reserves at the end of 2019 were 1,069,636 million tons (Mt). Being an ancient energy source; coal production showed a slight increase (142.89–167.58 Mt) in the last decade. Carbon emissions in the last ten years increased by 1.1% yearly. According to the British petroleum survey, carbon emissions increased from 29,745.2 Mt to 34,169 Mt in a single decade.

Worldwide energy consumption has increased 17-fold in the last century, and emissions of CO_x, SO_x, and NO_x from fossil-fuel combustion are the primary cause of atmospheric pollution [6] and increased the GHGs [7]. Around 2 Mt of soot and dark particles are released annually only from the world's largest populations, which is responsible for heating the air and melting the glaciers. Therefore, the glaciers that provide water to south Asian nations are decreasing rapidly due to global warming resulting in catastrophic floods in the region. The fossil fuels beneath the earth's surface are not evenly distributed, promoting the search for alternative energy sources available globally [8]. Moreover, due to drastic climate changes and energy shortages, approaches to reduce environmental pollution and alternative energy resources are being explored [9–11].

Renewable energy production and consumption have been increased over time. Among the global renewable energy giants, China contributed 7.9% to the total renewable energy consumption in 2010, while in 2020, this share increased to 24.5% [12]. Besides the depletion of fossil fuels and environmental threats associated with their consumption, modern civilization has produced a tremendous amount of solid organic and inorganic wastes. The global solid waste generation rate was 0.3 Mt per day in 1900, which increased to 3 Mt per day in 2000, and it is supposed to be doubled by 2025. The world's largest landfills such as Laogang (China), Sudokwon (Seoul), the now-full Jardim Gramacho (Brazil), and Bordo Poniente (Mexico City) receive around 10,000 tons of waste daily [13]. In developed countries, the municipal solid waste generation was reported to be 1.43–2.08 kg/person/day; however, it was 0.3–1.44 kg/person/day for developing countries [14]. Solid waste (municipal solid waste) refers to any garbage, trash, or refuse material, which represents a potential cause of pollution.

IntechOpen

The risk of waste production is getting higher day by day, even in developing countries due to the increase in the world's population and urbanization [15]. Thus, it can be said that waste generation is directly proportional to the rate of population growth. Further, waste organic and inorganic components are equally important as they hold a potential threat to living organisms and the environment [16]. According to Environmental Protection Agency (EPA), solid waste could be hazardous or non-hazardous depending upon its source. It usually consists of everyday items that people throw away. Generally, it is characterized into two major types: trash and garbage/rubbish. Garbage can also refer to food waste or kitchen waste, comprising organic waste, clothing, and food containers. In contrast, trash consists of daily household or other items no longer needed, including furniture, leaves, grass clippings, and junk [17]. Other major waste classes include agricultural waste, bio-medical waste, chemical waste, radioactive waste, construction waste, and e-waste. Waste management is of prime focus worldwide as improper waste disposal has caused severe environmental issues such as air and water pollution, loss of endangered wildlife habitat, disease outbreaks, and climate change. All these have a direct impact on society as well as the world's economy. To treat waste properly, it is of utmost importance that waste is characterized and collected accordingly. In terms of municipal waste generation, the United States and Canada were the two of the largest per capita waste producers, generating almost 2.58 kg and 2.33 kg daily, respectively [18].

Organic waste has received great attention as it is biodegradable and can be broken down into methane, carbon dioxide, water, and other organic compounds. It could be in the form of food, green waste, or feces. Since the byproducts of organic waste are usually harmless, they can be used on an industrial scale to produce biofuels. Therefore, many countries are consuming waste to generate energy [19]. Organic waste could be the byproduct of various industries such as agriculture, meat, poultry, sugar refineries, and oil industries. The composition of organic waste constantly varies as it is a combination of a variety of compounds. It all depends upon the properties and amount of each component present in organic waste. Therefore, its characterization and segregation are equally crucial in extracting maximum nutrients cost-effectively [20, 21]. Studies have shown the importance of agricultural and livestock waste among organic wastes. With the increase in agro-based industrialization, waste production has been increased up to three folds. These residues are a rich source of biocompounds that can be used for biogas production and manufacturing enzymes, vitamins, antibiotics, and animal feed [22]. Agricultural and livestock waste is always preferred among the various types of organic waste [23, 24]. The waste of slaughterhouses and fallen stocks are also rich in organic compounds that can be converted into valuable biofuels [25]. Each year, more than 2 billion tons of agro-waste are piled up, comprising straw and husk of wheat, rice, and barley. Adding up to this is forest waste (0.2 billion cubic meters), municipal solid waste (1.7 billion tons), industrial waste (approximately 9 billion tons), and animal waste (1.3 billion tons) [26]. If the necessary measurements for waste treatment are not appropriately followed, society, humans, flora, and fauna will face many challenges. With the advancement in science and technology, scientists are focusing on the gross value of waste as the products of these waste treatments are aimed to be environmentally friendly. Organic wastes are considered a potential resource for several applications, including animal feed, raw material in different industries, and feedstocks for biofuel. The R&D for the utilization of various organic waste for biofuels including biodiesel [27], crude bio-oil [28], bioethanol [28], and biogas production [29] developed fast in the past decades due to its lower carbon and GHGs emissions and the reduction of toxic waste from the environment [30]. Among the various methods of using organic waste as an energy source, anaerobic digestion (AD) has gained the most attention.

2. Biogas production

Due to biogas production from organic waste in the last years, there is a relative decrease in greenhouse gas emissions and fossil fuel consumption. Biogas consists of 50–75% methane, 25–50% of carbon-dioxide, 1–2% ammonia, and traces of hydrogen sulfide, oxygen, nitrogen hydrogen, and fermented organic fertilizer [29]. Biogas generation is an economical method since the raw material primarily used is agricultural and food waste. It could also be termed green energy and can be used in boilers for heat generation [31]. The basic phenomenon of biogas is the conversion of solar energy stored in the organic waste into gaseous energy by anaerobic digestion. Therefore, biogas is generated by microorganisms as a byproduct of their metabolism. The total energy level could be calculated by methane quantity [32].

Various process variables affecting biogas production, like the nature of the feedstock and carbon-to-nitrogen ratio, and reactors setup, have been evaluated. Different agricultural residues (wheat stalk, soybean straw, and black gram stalk), food wastes, and animal wastes are suitable for biogas production [31, 33, 34]. In recent years, it has been observed that landfills for waste management had specific side effects on the environment. Previously, biogas plants were established for waste disposals. Nevertheless, this practice has been changed ever since. These plants are now used for energy generation from biomass. For this purpose, many studies have been conducted to evaluate the optimal capacity of waste being converted into energy with greater yields and cost-effective mechanisms [19]. Biogas is used as fuel on the domestic and commercial levels. The production capacity from the installed biogas plants across the globe has been increasing every year.

As a renewable energy, biomethane can be derived from various substrates under anaerobic conditions, including sewage and waste activated sludge, food wastes and vegetable, wastes from forestry, manure from living stocks, agriculture wastes, and wastewater [35]. Biogas derived from organic wastes through AD is suitable to clean energy to fulfill energy demand [36]. AD is commonly considered a reliable and cheap approach for energy recovery and wastes management [37], which minimizes the waste quantity and uncontrolled emissions. Besides, the AD digestates contain nutrients and can serve as a biofertilizer for crops. Biogas might substitute fossil fuels and lower the GHGs emission at households and commercial scale [38]. AD of different feedstocks may have a different biomethane production and obtain more bioenergy to compensate for the net energy utilized during the process.

3. Feedstocks for biogas generation

Most of the biowaste is landfilled, burned, or only reused after composting. However, it can be utilized as a potential source of bioenergy through different practices [10]. A variety of biowaste can be used as substrate in AD to generate clean and renewable energy in the form of biogas and biomethane [39, 40]. Biowaste is mainly composed of 3 major biocomponents, i.e., lipids, proteins, and carbohydrates. Agricultural waste, forest waste, wood residues, fruit and vegetable waste, and municipal sludge contains high content of carbohydrates-based compounds. Protein biowaste is mainly originated from animal sources such as slaughterhouse waste, meat processing industries, and dairy industries. The lipids-based feedstocks are derived from waste oil, oil mills, animal fats from the slaughterhouse, FOG, grease trap waste from sanitation, and wastewater from restaurants. Most of this waste has been applied in AD to generate biogas [7, 41]. Among carbohydrates, lipids, and proteins, the maximum biogas production potential has been reported for lipids. The energy potential of organic wastes and biomass mostly depends on their physiochemical and elemental commotions [42]. Among which volatile solids (VS) and the ratio of carbon to nitrogen (C/N ratio) are the most important as only the organic portion in any waste is attributed as VS, and the carbon is used as food during the microbial process to produce bioenergy [43]. Moisture is another essential aspect for improved degradability of biomasses, especially in agriculture, fruits, and vegetable waste [44]. The COD (chemical oxygen demand) of organic waste material corresponds to the amount of organic substrate available to the microbial community for biogas production [45].

The present book aims to discuss biogas production from different resources and the impact and changes of microbial community during the digestion process. In addition, the possible utilization of biogas byproducts as biofertilizers will be evaluated. Moreover, case studies on biogas production from municipal solid wastes will be presented.

Author details

El-Sayed Salama[1] and Abd El-Fatah Abomohra[2*]

1 Department of Occupational and Environmental Health, School of Public Health, Lanzhou University, Lanzhou, China

2 Department of Environmental Engineering, School of Architecture and Civil Engineering, Chengdu University, Chengdu, China

*Address all correspondence to: abomohra@cdu.edu.cn

References

[1] Gorus MS, Ben Ali MS. Globalization and environmental pollution: Where does the MENA region stand? In: Ali MSB, editor. Economic Development in the MENA Region: New Perspectives. Cham: Springer International Publishing; 2021. pp. 161-179

[2] Pata UK. Linking renewable energy, globalization, agriculture, CO_2 emissions and ecological footprint in BRIC countries: A sustainability perspective. Renewable Energy. 2021;**173**:197-208

[3] Abomohra AE-F, Elsayed M, Esakkimuthu S, El-Sheekh M, Hanelt D. Potential of fat, oil and grease (FOG) for biodiesel production: A critical review on the recent progress and future perspectives. Progress in Energy and Combustion Science. 2020;**81**:100868

[4] Hosseinpour S, Aghbashlo M, Tabatabaei M. Biomass higher heating value (HHV) modeling on the basis of proximate analysis using iterative network-based fuzzy partial least squares coupled with principle component analysis (PCA-INFPLS). Fuel. 2018;**222**:1-10

[5] BP. BP Statistical Review of World Energy. 69th ed. England: BP International Limited; 2020. Available from: https://www.bp.com/

[6] Yin J, Ding Q, Fan X. Direct and indirect contributions of energy consumption structure to carbon emission intensity. International Journal of Energy Sector Management. 2021;**15**(3):665-677

[7] Omer AM. An overview of biomass and biogas for energy generation: Recent development and perspectives. Agricultural Advances. 2016;**5**(2):251-268

[8] Laghari JR. Melting glaciers bring energy uncertainty. Nature. 2013; **502**(7473):617-618

[9] Abomohra J, Tu R, Han SF, Eid M, Eladel H. Microalgal biomass production as a sustainable feedstock for biodiesel: Current status and perspectives. Renewable and Sustainable Energy Reviews. 2016;**64**:596-606

[10] Almomani F, Bhosale R, Khraisheh M, Shawaqfah M. Enhancement of biogas production from agricultural wastes via pre-treatment with advanced oxidation processes. Fuel. 2019;**253**:964-974

[11] Tabatabaei M, Aghbashlo M, Valijanian E, Panahi HKS, Nizami A-S, Ghanavati H, et al. A comprehensive review on recent biological innovations to improve biogas production, part 1: Upstream strategies. Renewable Energy. 2020;**146**:1204-1220

[12] Bernard L. Statistical Review of World Energy. England: BP International Limited; 2021. Available from: https://www.bp.com/

[13] Hoornweg D, Bhada-Tata P, Kennedy C. Waste production must peak this century. Nature. 2013; **502**(7473):615-617

[14] Troschinetz AM, Mihelcic JR. Sustainable recycling of municipal solid waste in developing countries. Waste Management. 2009;**29**(2):915-923

[15] Guerrero LA, Maas G, Hogland W. Solid waste management challenges for cities in developing countries. Waste Management. 2013;**33**(1):220-232

[16] Balyani V, Sharma S, Goyal A. A Review on Sustainable Solid Waste Management for UT: Case Study of Chandigarh city. International Research Journal of Engineering and Technology. 2021;**8**:1306-1310

[17] Vergara SE, Tchobanoglous G. Municipal solid waste and the

environment: A global perspective. Annual Review of Environment and Resources. 2012;**37**:277-310

[18] Tiseo I. Global per capita generation of municipal solid waste by select country 2018. 2020. Available from: https://www.statista.com/statistics/689809/per-capital-msw-generation-by-country-worldwide

[19] Szabó E, Nagy VJS. Biogas from organic wastes. 2011;**21**(4):887-891

[20] Abomohra AE-F, Wang Q, Huang J. Waste-to-Energy: Recent Developments and Future Perspectives towards Circular Economy. Germany: Springer Nature; 2022

[21] Shareef TME, Zhao B. The fundamentals of biochar as a soil amendment tool and management in agriculture scope: An overview for farmers and gardeners. Journal of Agricultural Chemistry and Environment. 2016;**6**(1):38-61

[22] Sadh PK, Duhan S, Duhan JS. Agro-industrial wastes and their utilization using solid state fermentation: A review. Bioresources and Bioprocessing. 2018;**5**. DOI: 10.1186/s40643-017-0187-z

[23] Sharma K, Garg V. Vermicomposting of waste: A zero-waste approach for waste management. In: Sustainable Resource Recovery and Zero Waste Approaches. Singapore: Elsevier; 2019. pp. 133-164

[24] Tang J, Pu Y, Zeng T, Hu Y, Huang J, Pan S, et al. Enhanced methane production coupled with livestock wastewater treatment using anaerobic membrane bioreactor: Performance and membrane filtration properties. Bioresources Technology. 2022;**345**:126470

[25] Lee J, Cho S, Kim D, Ryu J, Lee K, Chung H, et al. Conversion of Slaughterhouse Wastes to Solid Fuel Using Hydrothermal Carbonization. Energies. 2021;**14**(6):1768. DOI: 10.3390/en14061768

[26] Millati R, Cahyono RB, Ariyanto T, Azzahrani IN, Putri RU, Taherzadeh MJ. Agricultural, industrial, municipal, and forest wastes: An Overview. In: Taherzadeh M, Bolton K, Wong J, Pandey A, editors. Sustainable Resource Recovery and Zero Waste Approaches. Singapore: Elsevier; 2019. pp. 1-22

[27] Almutairi AW, Al-Hasawi ZM, Abomohra AE. Valorization of lipidic food waste for enhanced biodiesel recovery through two-step conversion: A novel microalgae-integrated approach. Bioresources Technology. 2021;**342**:125966

[28] Wang S, Hu S, Shang H, Barati B, Gong X, Hu X, et al. Study on the co-operative effect of kitchen wastewater for harvest and enhanced pyrolysis of microalgae. Bioresource Technology. 2020;**317**:123983

[29] Atelge MR, Krisa D, Kumar G, Eskicioglu C, Nguyen DD, Chang SW, et al. Biogas production from organic waste: Recent progress and perspectives. Waste and Biomass Valorization. 2020;**11**(3):1019-1040

[30] Dahunsi SO, Adesulu-Dahunsi AT, Osueke CO, Lawal AI, Olayanju TMA, Ojediran JO, et al. Biogas generation from Sorghum bicolor stalk: Effect of pretreatment methods and economic feasibility. Energy Reports. 2019;**5**:584-593

[31] Sindhu R, Binod P, Pandey A, Ankaram S, Duan Y, Awasthi MK. Chapter 5—Biofuel production from biomass: Toward sustainable development. In: Kumar S, Kumar R, Pandey A, editors. Current Developments in Biotechnology and Bioengineering. Singapore: Elsevier; 2019. pp. 79-92

[32] Al-Wahaibi A, Osman AI, Al-Muhtaseb AH, Alqaisi O, Baawain M, Fawzy S, et al. Techno-economic evaluation of biogas production from food waste via anaerobic digestion. Scientific Reports. 2020;**10**:15719. DOI: 10.1038/s41598-020-72897-5

[33] Elsayed M, Abomohra AE, Ai P, Wang D, El-Mashad HM, Zhang Y. Biorefining of rice straw by sequential fermentation and anaerobic digestion for bioethanol and/or biomethane production: Comparison of structural properties and energy output. Bioresources Technology. 2018; **268**:183-189

[34] Usman M, Zha LJ, Abomohrad A, Li XK, Zhang CJ, Salama E. Evaluation of animal- and plant-based lipidic waste in anaerobic digestion: Kinetics of long-chain fatty acids degradation. Critical Reviews in Biotechnology. 2020;**40**(6):733-749

[35] Watson J, Zhang Y, Si B, Chen W-T, de Souza R. Gasification of biowaste: A critical review and outlooks. Renewable and Sustainable Energy Reviews. 2018;**83**:1-17

[36] Aghbashlo M, Tabatabaei M, Soltanian S, Ghanavati H, Dadak A. Comprehensive exergoeconomic analysis of a municipal solid waste digestion plant equipped with a biogas genset. Waste Management. 2019;**87**:485-498

[37] Kothari R, Pandey A, Kumar S, Tyagi V, Tyagi S. Different aspects of dry anaerobic digestion for bio-energy: An overview. Renewable and Sustainable Energy Reviews. 2014;**39**:174-195

[38] Yıldırım E, Ince O, Aydin S, Ince B. Improvement of biogas potential of anaerobic digesters using rumen fungi. Renewable Energy. 2017;**109**:346-353

[39] Salama E-S, Saha S, Kurade MB, Dev S, Chang SW, Jeon B-H. Recent trends in anaerobic co-digestion: Fat, oil, and grease (FOG) for enhanced biomethanation. Progress in Energy Combustion Science. 2019;**70**:22-42

[40] Tabatabaei M, Aghbashlo M, Valijanian E, Panahi HKS, Nizami A-S, Ghanavati H, et al. A comprehensive review on recent biological innovations to improve biogas production, Part 2: Mainstream and downstream strategies. Renewable Energy. 2020b;**146**:1392-1407

[41] Busic A, Kundas S, Morzak G, Belskaya H, Mardetko N, Santek MI, et al. Recent trends in biodiesel and biogas production. Food Technology and Biotechnology. 2018;**56**(2):152-173

[42] Arif M, Li YX, El-Dalatony MM, Zhang CJ, Li XK, Salama E. A complete characterization of microalgal biomass through FTIR/TGA/CHNS analysis: An approach for biofuel generation and nutrients removal. Renewable Energy. 2021;**163**:1973-1982

[43] Panizio RM, Calado LFC, Lourinho G, de Brito PSD, Mees JB. Potential of biogas production in anaerobic co-digestion of opuntia ficus-indicaand slaughterhouse wastes. Waste and Biomass Valorization. 2020;**11**(9):4639-4647

[44] Scano EA, Asquer C, Pistis A, Ortu L, Demontis V, Cocco D. Biogas from anaerobic digestion of fruit and vegetable wastes: Experimental results on pilot-scale and preliminary performance evaluation of a full-scale power plant. Energy Conversion and Management. 2014;**77**:22-30

[45] Duong TH, Grolle K, Nga TTV, Zeeman G, Temmink H, van Eekert M. Protein hydrolysis and fermentation under methanogenic and acidifying conditions. Biotechnology Biofuels. 2019;**12**:254

Chapter 2

Biogas Production: Evaluation and Possible Applications

Venko Beschkov

Abstract

Biogas is an excellent example of renewable feedstock for energy production enabling closure of the carbon cycle by photosynthesis of the existing vegetation, without charging the atmosphere with excessive carbon dioxide. The present review contains traditional as well as new methods for the preparation of raw materials for biogas production. These methods are compared by the biogas yield and biogas content with the possible applications. Various fields of biogas utilization are discussed. They are listed from simple heating, electricity production by co-generation, fuel cell applications to catalytic conversions for light fuel production by the Fischer-Tropsch process. The aspects of carbon dioxide recycling reaching methane production are considered too.

Keywords: biogas, raw materials, pre-treatment, production, utilization

1. Introduction

The extensive economic growth in the developed countries imposed a severe impact on the air and water quality. The impact on the air quality consists in the enormous emissions of greenhouse gases from many sources: energy production, burning fossil fuels, transport and household. The emission rate of resulting carbon dioxide in the atmosphere is too high to enable its assimilation by the present vegetation. Therefore, the concept of the use of renewable energy sources became so important during the last decades. Besides wind and solar energy, an important place occupies the biomass, namely biogas, bioethanol, biodiesel. The main reason is the replacement of fossil fuels by carbon-containing biomass that can be easily assimilated by the present vegetation. Hence, the carbon cycle is closed. Another option of use of biomass is its application as raw material for chemical productions thus replacing, at least partially the oil as the main feedstock for organic synthetic products [1–3]

Biogas is the simplest renewable fuel in comparison with bioethanol and biodiesel. Besides its use as a fuel, it can be converted into other products, like light fuels and chemical products after dry reforming and consequent catalytic conversions, like the Fischer-Tropsch process [4]. Many countries adopted programs for biogas applications in energy production [5, 6]. The comparison between the biofuels produced by different biomass as the substrate is shown below.

Biogas is produced by anaerobic digestion of organic materials from the natural origin [7–9]. Normally it contains methane (50–75% vol.). The rest is carbon dioxide with small amounts of nitrogen, hydrogen, ethane and traces of sulfur compounds (hydrogen sulfide and mecaptanes). The calorific value of biogas is between 20 and 32 MJ/m^3. Its production became popular in the first half of the twentieth century. It became more important after the growth of the oil and gas prices in the 1970s. On the other hand, its importance is steadily maintained in the developing countries in Asia and Africa where a lot of low-scale anaerobic digesters are developed and used in the household [10, 11]. Currently, biogas production is popular in Europe and North America as a tool for simultaneous treatment of waste in agriculture and food industry and energy production at the same time to maintain these activities [10, 12, 13]. Landfill gases containing methane are also the reason for concern, since the emitted methane has a 25 times stronger greenhouse effect than carbon dioxide. There are also practical applications for the utilization of these gases for energy production thus reducing their harmful greenhouse effect.

The global energy production from biogas in the year 2000 was about 280,000 TJ, growing to almost 1.3 million TJ by 2014. As a volume, the annual world production of biogas was about 59 billion cubic meters in 2013. Almost half of this amount was produced in the European Union [10, 12, 14, 15] and it is growing considerably during the last decade [13].

The classical substrate for biogas production is manure (cattle, pig), poultry litter and activated sludge. However, there are other carbon sources to be treated by anaerobic digestion, like lingo-cellulosic residues, waste from the food industry, like stillage from ethanol distilleries, vegetable and meat industries, etc. In some cases, these substrates must be pre-treated to be converted into digestible form [14–16].

In the present chapter different substrates for biogas production will be considered along with their pre-treatment and mode of operation. Different applications of biogas will be outlined below.

2. Substrates and biogas yields

The traditional substrate for biogas production is manure, poultry litter, lingocellulose, activated sludge, as well as residues from the food industry (stillage from alcohol beverage production, vegetable waste, etc.). The gas yield per unit mass of substrate is an important indicator for further decisions for process development and plant construction. There are various data for this indicator but here we shall present some of them as average figures, cf. **Table 1** [17].

The best methane yield per unit of total solids can be attained by grass as a substrate. In general, the choice of substrate depends on various factors: its availability and the problems it may cause; the economic issues, as the price of energy and waste treatment; the equipment for anaerobic digestion, etc. For example, the

Substrate	Biogas yield, m^3/kg TS	Methane content, %vol.	Specific methane yield, m^3/kg TS	Reference
Cattle manure	0.29	62.8	0.182	[16]
Pig manure	0.43	66.9	0.288	[17, 18]
Poultry litter	0.47	57.9	0.277	[19]
Grass	0.55	77	0.423	[17]
Straw	0.34	58.0	0.197	[17]
Corn stalks	0.42	53.0	0.226	[17]

Table 1.
Experimentally estimated biogas yields per unit total solids (TS) from different agricultural waste.

use of grass gives the best results but requires additional pre-treatment to facilitate the conversion of non-soluble lignocellulose into soluble and biodegradable oligosaccharides. More detailed survey on the waste potential for biogas production is given in [17–20].

3. Pretreatment of substrates

The pretreatment methods of biomass for biogas production depend on the type of substrate and it is associated with the main scheme of consecutive steps of AD. [21].

The pretreatment method is related closely to the first step in the technology, i.e., the hydrolysis of insoluble organics. As a result, the macromolecules, i.e., carbohydrates, proteins, lipids are converted into soluble and digestible compounds of lower molecular mass.

The main groups of pretreatment methods of organic substrates for biogas production are mechanical, chemical methods [22] and microbial ones [8].

Milling is an inevitable step in substrate pretreatment reducing the size of the material particles. It can improve susceptibility to enzymatic hydrolysis of lignocelluloses [23].

3.1 Chemical methods

The chemical methods are based on acid or alkaline hydrolysis of the natural polymers.

The biggest problems are met with the pre-treatment of lingo-cellulosic substrates. The main problem is the removal of lignin. Alkaline hydrolysis is used for this purpose. Sodium hydroxide, lime or ammonia are applied with a substantial increase of biogas production, up to 16% vol. [24, 25].

Another chemical method is the treatment of substrates by calcium hypochlorite, combining chemical oxidation with alkaline action. There are new data for the treatment of waste-activated sludge by $Ca(ClO)_2$ thus increasing the methane yield up to 60% [26].

The acid hydrolysis of ligno-cellulose substrates consists of the treatment of the substrate by sulfuric acid [27], but hydrochloric acid and nitric acid also have been used [22]. The acid hydrolysis is usually accomplished at higher temperatures (120-180°C) and pressure. Under these conditions hemicellulose is completely degraded, cellulose to a higher extent. However, lignin is only partially degraded.

A serious disadvantage of these two kinds of chemical treatment is the necessity of pH adjustment because of the sensitivity of the methanogenics toward pH. It is known they can successfully produce methane in the pH range of 6–8.

Best results of chemical treatment are obtained by H_2O_2 in alkaline media combined with microwave treatment [28]. However, the price of H_2O_2 makes this method unpractical.

There are also some efforts for pretreatment by ozonolysis [29, 30], ionic liquids [31–33]. But they are too costly for large-scale practical application.

3.2 Thermal methods and steam explosion

3.2.1 Thermal pretreatment

Besides the thermochemical methods (acid and alkaline hydrolysis) thermal pretreatment consist of purely thermal treatment. First, it is treatment by hot water at elevated pressure so keep water at liquid state [34].

This kind of pretreatment facilitates the further enzymatic digestibility of cellulose with better sugar yield and almost no fermentation inhibitor [35]. An advantage of this method is the lack of chemicals and additional waste streams and it is eco-friendly because it does not need neutralization of liquid streams and conditioning chemicals saving time and energy for it.

A number of chemical reactions takes place during hot water pretreatment. The thermal destruction of hemicellulose results in the production of organic acids. They act as catalysts to promote the hydrolysis of carbohydrate polysaccharides into oligosaccharides and monosaccharides. These processes resemble dilute acid hydrolysis.

3.2.2 Steam explosion

The method of steam explosion consists of the action of saturated steam at high pressure on the biomass for some time. Afterwards the pressure is released abruptly causing the explosive breakdown of the macromolecules in the biomass and the bonds between them [36]. It is a widely used method of biomass pretreatment for various purposes (ethanol fermentation, biogas production, etc.). It is considered a catalyzed and uncatalyzed steam explosion. In the first case, some acidic chemicals (SO_2, H_2SO_4, CO_2) are used as catalysts to mix with biomass before steam-explosion. The commonly used temperature range is 160–260°C for short period of time at pressures up to 4.8 MPa [37].

During uncatalyzed steam, explosion hemicellulose is degraded and lignin structure is altered. The cellulose digestibility during steam explosion followed by enzymatic hydrolysis is enhanced [38, 39]. However, the catalyzed steam explosion is considered more efficient because of the deeper transformation of the biomass into mode digestible intermediates but in some cases neutralization of the mixture is required, e.g., when sulfuric acid is applied.

Certain limitations associated with the steam explosion method are: (1) incomplete disruption of fibers, (2) generation of inhibitory components to microbial growth, enzymatic hydrolysis and fermentation [40]. Because inhibitory degradation products are formed pretreated biomass needs to be washed with water to remove the inhibitory materials along with water-soluble hemicelluloses [41]. The apparent increase of lignin content during heat treatment has been observed due to hemicelluloses degradation product, furfural and lignin polymerization [41].

The more profound removal of lignin is a key-step in biomass pre-treatment before anaerobic digestion and therefore special attention is paid to it, cf. Timilsena [42].

3.2.3 Enzyme methods

This group of methods is essential for biomass pre-treatment. It can be applied in combination with other ones, as mentioned above or separately. Usually, it is relied on enzymes existing in the very biomass, for example in cattle manure [43]. Otherwise, isolation and application of certain hydrolases for the aims of biogas production are not economically acceptable.

The main microbial species, capable to convert the insoluble substrates into soluble ones are from the genera *Pseudomonas, Cellulomonas, Streptomyces, Bacillus*, etc. and white-rot fungi (like *Trichoderma, Aspergillus, Penicillium*) as well [9].

There are data about the capability of certain fungi to degrade lignin, thus enabling further cellulose hydrolysis, see Lee et al. [44].

Anyway, the enzyme methods are naturally incorporated into the overall hydrolytic process of biomass preparation for further acidogenesis, cf. **Figure 1**.

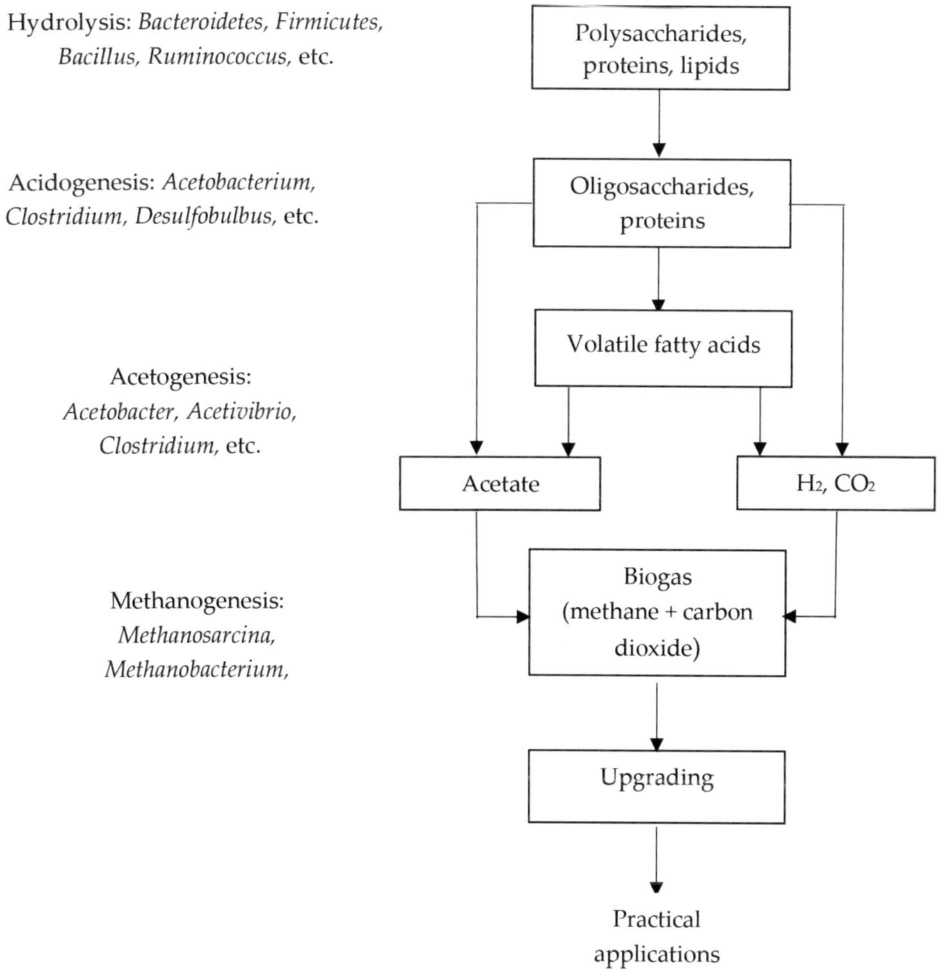

Figure 1.
Scheme of the consecutive processes of biogas formation in anaerobic process, according to Garcia-Heras [21].

3.2.4 Other methods

There are also some physical methods, including pretreatment by γ-irradiation [45], by ultrasonication [46], pulsed electric field [47] with electric field intensity of up to 20 kV/cm. The main disadvantage of these methods is that they are high energy-consuming and therefore very costly.

Microwave treatment has been also considered [48–50]. Our experience with microwave treatment of corn stalks and grass hey did not give better results for biogas yield compared to the treatment by acid hydrolysis or simple enzyme treatment.

Recently a constant electric field was applied after the steam explosion of activated sludge to remove or destroy the inhibitors formed during the steam explosion. A very high methane yield was observed. The same approach was also applied to other substrates, like cattle manure, coniferous needles, glycerol and their mixtures [51]. Some results are shown below, cf. **Figure 2**.

In the conducted experiments, we have found out that, the treatment of the waste material with electric current leads to improvement in the ingredients of produced biogas, expressed mainly in higher methane content (reaching 95–98% (vol.) in experiment E4 in a comparison with most commonly observed 50–75%).

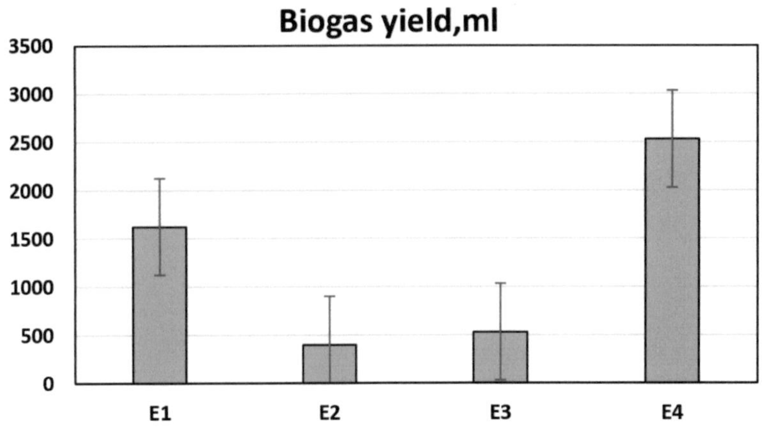

Figure 2.
Cumulative biogas yield for ca. 110 days under different pre-treatment conditions at different amounts of added glycerol, manure and sulfuric acid at different anode potential. Mesophilic process..E1–16 g coniferous material +200 ml 1% H_2SO_4 *+ 8 g glycerol +600 g manure. Treated by constant anode potential 0.77 V/S.H.E. for 30 minutes. E2–32 g coniferous material +400 ml 1%* H_2SO_4 *+ 16 g glycerol +1200 g manure. Treated by constant anode potential 0.77 V/S.H.E. for 30 minutes. E3–16 g coniferous material +100 ml 1%* H_2SO_4 *+ 8 g glycerol +300 g manure. Treated by constant anode potential 0.25 V/ S.H.E. for 30 minutes. E4–16 g coniferous material + + 100 ml 1%* H_2SO_4 *+ 8 g glycerol +300 g manure. Treated by constant anode potential 0.5 V/ S.H. E. for 30 minutes. Original data reported in [51].*

One can see that moderate amounts of glycerol and manure with the low amount of sulfuric acid are preferable (experiment E4). We have observed that applying electrical current to cattle manure leads to the intensification of the digesting process, more biogas and higher methane content. The importance of the anode potential is visible after a comparison of the results from experiments E3 and E4. Under similar initial components of the reactive mixtures, the anode potential of 0.5 V/S.H.E. is superior to the one at E3, namely 0.25 V/S.H.E.

Generally, the decision for selection of the certain method of pretreatment has its technological and economic backgrounds and the cheapest one must be chosen depending on the very conditions. For example, simple microbial hydrolysis by cellulases contained in the cattle manure could be sufficiently effective compared to the sophisticated physical and thermochemical processes.

4. Biogas production

The mesophilic anaerobic digestion with biogas production follows the steps described in **Figure 1**.

The operation conditions for the production of biogas are associated with the selected substrate. Generally, the first choice is to decide whether the process will be mesophilic (30-40°C) or thermophilic one (50–60°C) [9].

The thermophilic process seems to be preferable because of the higher biogas production rate. Another reason is the sterilization of the sludge destroying pathogenes and parasite microbial cultures. Next, undesirable seeds of various weeds contained in the manure are also destroyed thus protecting the soil from weeds at further fertilization by the residual sludge and wastewater. However, the thermal balance of the produced energy and the energy input to maintain a higher temperature must be made carefully. Another unexpected obstacle is the higher sensitivity of the thermophilic microbes to pH variation than the mesophilic ones. From this point of view, the mesophilic process seems to be more promising.

The effectiveness of microbes involved in anaerobic digestion determines the rate of substrate decomposition and biogas production [52]. The naturally formed microbial consortia are quite sensitive to pH variations and unbalanced operation may lead to strong inhibition and process failure. At mesophilic processes, the hydrolysis is usually performed by bacteria from the genera *Bacillus, Streptococcus, Klebsiella*, etc. After hydrolysis, the following acidogenesis, acetogenesis and methanization take place, performed by different specific bacteria and archae.

Acidification is usually performed by bacteria from the genera *Acetobacterium, Clostridium, Desulfobulbus, Eubacterium,* etc. [52]. At a higher feeding rate of the substrate in the acidification phase, an excessive production of volatile fatty acids (formic, acetic, propionic, butyric ones) may occur thus decreasing pH below the optimum value for methanogens (i.e., between 7 and 8). It usually provokes stopping the process of biogas production. High volatile fatty acid (VFA) concentrations inhibit the growth of acid-producing bacteria thus reducing also rate of acidogensis. Fermentation of sugar is inhibited by total concentrations of volatile fatty acids above 4 g/l [15, 53]. The long-chain fatty acid concentration of about 30–300 mg/l was found as appropriate for anaerobic decomposition [9]. Some of the possible acidogenic reactions are listed below.

$$C_6H_{12}O_6 + 2H_2 = 2CH_3CH_2COOH + 2H_2O \quad (1)$$

$$C_6H_{12}O_6 + 2H_2O = 2CH_3COOH + 2CO_2 + 4H_2 \quad (2)$$

The last one is an acetogenic one too. Formic acid is formed by acetogenesis and it is carried out by bacteria from the genera *Syntrophomonas, Syntrophobacter, Clostridium, Syntrophospora, Acetobacter* [9]. Acetate is formed from propionate, bicarbonate too:

$$CH_3CH_2COO^- + 3H_2O = CH_3COOH + HCO_3^- + 3H_2 \quad (3)$$

$$2HCO_3^- + 4H_2 + H^+ = CH_3COO^- + 4H_2O \quad (4)$$

Further VFA is decomposed to methane and carbon dioxide following the reactions (5):

$$\begin{aligned} 4HCOOH &= CH_4 + 3CO_2 + 2H_2O \ (\textit{Methanobrevibacter}, \textit{Methanococcus}) \\ CH_3COOH &= CH_4 + CO_2 \ (\textit{Methanosarcina}) \\ C_2H_5COOH + 2H_2O &= CH_4 + 2CO_2 + 3H_2 \end{aligned} \quad (5)$$

Other schemes for methane formation from carbon dioxide and hydrogen is accomplished by *Methanobacterium, Methanobrevibacter, Methanothermus* [54]:

$$CO_2 + 4H_2 = CH_4 + 2H_2O \quad (6)$$

In this case the content of methane in the biogas is much higher. A molar (or volumetric yield) of biogas, richer of methane more than 50% is a clear indication for the pathway, shown in Eq. (6).

The method of carbon isotopes was extensively used to establish the pathway of methane production, as summarized by Conrad [55].

4.1 Single stage and multi-stage systems

It is well, in any case, to manage biogas production in a continuous or fed batch mode. The latter is preferable because of the low process rate and the menace of wash-out (at high dilution rates) or accumulation of inhibitors (at high substrate dosage).

The anaerobic digestion systems are classified as single stage and multi-stage systems based on the different steps of digestion, cf. **Figure 1**. The simple single stage systems are used for many decades. They are available in the simplest design. The most used type of anaerobic digestion is the UASB (upflow anaerobic sludge blanket) digester used extensively for wastewater treatment and biogas production [56].

The substrate is introduced in the lower end of the bioreactor and passed through a layer of sludge, where granules of microbes are formed. The treated water leaves the digester from the upper side where the produced biogas is also released.

All phases of anaerobic digestion are carried out in a single apparatus with batch or continuous mode. In a single digester, all steps of anaerobic digestion take place in one space and as a result process fluctuations or low biogas production occurs because of the accumulation of inhibitors (volatile fatty acids or ammonia in some cases) at hydrolysis and acidogenesis. As a result, the pH in the digester goes out of the optimum limits for successful methanogenesis. Better and more stable operation is possible when the steps in **Figure 1** are carried out simultaneously, but separated in consecutively situated reactors. Multi-stage cascades of bioreactors with the separated acidogenesis and methanogenesis show better gas production. The main advantage of the consecutive scheme consists in the higher stability of each unit in the cascade at the undesired fluctuation of feed, substrate content, pH, temperature, etc. That is why the separation of each of the four stages of biogas production will improve substrate degradation as well as biogas/methane production. This approach was proposed by Grobicki & Stuckey [57] and later applied in a series of studies on stillage conversion to biogas [58] and glycerol utilization [59]. Microbial analysis showed that different microbial cultures were developed in the different steps, corresponding to the content of volatile fatty acids in the step [58, 60].

4.2 Biogas production with glycerol addition

Crude glycerol is the main waste product from biodiesel manufacturing. It is released in the amount equivalent to the methanol used and exceeds the market demands. This waste product contains water and it is contaminated by the catalyst and residual methanol. The demand for pure glycerol and its price makes the purification of this waste product not economically feasible. That is why the use of this glycerol for the production of value-added chemicals was sought [61–63]. Such products are propylene glycol, 1,3-propanediol, epichlorohydrin. Some of its derivatives are suitable as additives to gasoline and diesel.

There are some efforts for waste glycerol utilization as a substrate for biogas production [64–66]. There is also a study on glycerol addition to enhance the mutual production of biohydrogen and methane by crude glycerol addition [67]. It was established that glycerol and microalgal biomass as co-substrates had an antagonistic effect on hydrogen production and a synergistic effect on methane fermentation.

A hinder for this application is the rapid accumulation of VFA leading to strong inhibition of the methanogenesis and shift to production of gas with very low methane content [64, 65]. It is because in comparison to the traditional substrates glycerol has a very simple molecule and therefore it quickly yields intermediates and final products as organic acids and alcohols. If the initial amount of glycerol is high, the resulting pH drop leads to inhibition of methanogenesis. However, that small amounts of glycerol can boost biogas production based on traditional substrates, see Wohlgemut [68] and Fountoulakis & Manios [69].

When the digestate of bioethanol production was supplemented with 15% and 25% g/L of glycerol (as COD), the cumulative methane and biogas yield was increased to 318 Nml/gCOD and 196 Nml/gCOD which was approximately 6 times higher compared to digestion of the single substrate [66].

In our studies, we have shown that besides the biogas production some other valuable chemical products are obtained (2,3-butanediol, 1,3-propanediol) [60].

A multi-stage cascade bioreactor of eight consecutive compartments was used for anaerobic digestion of stillage with small controlled amounts of glycerol. The latter has been added in a fed-batch mode.

A specific microbial profile is formed along with the compartments, cf. **Table 2**. The bacteria of the strain *Klebsiella* are capable to digest glycerol to 1,3-propane diol and 2,3-butanediol. They can also produce formic acid to yield carbon dioxide and hydrogen. The microbial analysis showed that methane was produced mostly by the pathway of CO_2 reduction by hydrogen, cf. Eq. (7). It is also seen that the methanogens prevail in the second compartment and further.

In the next **Figure 3** the VFA profile, the pH profile and the concentrations of 2,3-butanediol along the bioreactor compartments are shown, see [60]. The VFA

Compartment No.	Genera	Pathway of methane production
1	Molds, *Bacillus*	None
2	*Klebsiella, Methanosarcina*	Acetate, Eq. (5); $CO_2 + H_2$, Eq.(6)
3	*Klebsiella, Methanobacterium*	$CO_2 + H_2$, Eq.(6)
4	*Klebsiella, Methanobacterium*	$CO_2 + H_2$, Eq.(6)
5	*Klebsiella, Methanobacterium*	$CO_2 + H_2$, Eq.(6)
6	*Klebsiella, Methanobrevibacter*	$CO_2 + H_2$, Eq.(6)
7	*Klebsiella, Methanobrevibacter*	$CO_2 + H_2$, Eq.(6)
8	*Klebsiella Methanobrevibacter*	$CO_2 + H_2$, Eq.(6)

Table 2.
Microbial profile in a multistage bioreactor with glycerol as a supplement. Microbial identification is taken from [60].

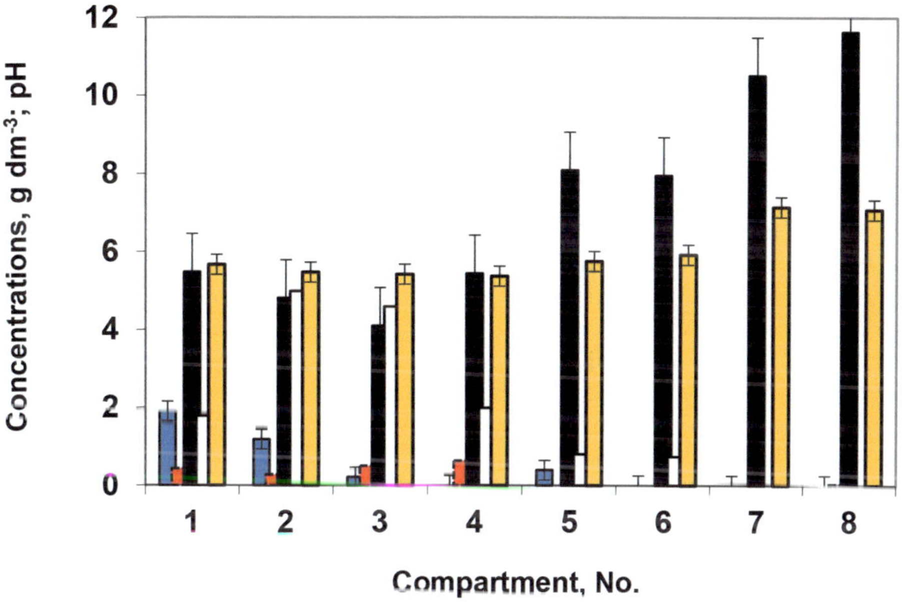

Figure 3.
Profiles of the substrate, intermediate products and pH on the 12th day after a feed with glycerol; (blue) – Glycerol; (red) – Acetic acid; (white) – Propionic acid; (black) – 2,3-butanediol; (yellow) – pH. Feed 1 kg crude glycerol, cf. ref. [60].

concentrations reach maximum values in compartments 2 and 3 and decrease along the bioreactor to zero in compartments 7 and 8. Obviously, acetic acid is converted more rapidly than the propionic one. The pH profile along the compartments correlates reasonably with the VFA variations.

The target product, i.e., 2,3-butanediol is accumulated in practically interesting concentrations, up to 12 g/l.

4.3 Biogas applications

The first and the simplest mode of application of biogas is direct combustion for heating and lighting, as it was adopted in developing countries. The next more sophisticated application is its use for power generation, so-called co-generation. Co-generation is the simultaneous production of electricity and heat by the combustion of biogas. This is the so-called "combined heat and power" process (CHP). Such applications are well spread, using municipal solid waste [70], activated sludge from wastewater treatment plants. The co-generation unit is composed of an engine that actuates as an alternator. The electricity efficiency of the co-generation units reaches 35%. The heat recovery makes it possible to reach a total output of 85% of all produced heat is utilized [71].

The flexibility of biogas systems can support electricity production to follow the temporal local electricity demand, thus facilitating grid stability. It is a decentralized component of the overall energy system and it can serve as a distribution hub in rural areas [72].

Besides CHP, another approach is the "power to gas" (PtG) where the surplus renewable electricity is used for the production of hydrogen by electrolysis [72]. It was proposed for utilization of energy surplus produced by traditional power stations when the electricity demand is low. There are proposals for carbon dioxide recycling by using the released hydrogen for the reduction of carbon dioxide to methane.

Biogas has already broad applications in transport. After upgrading, i.e., separation of carbon dioxide and the sulfur-containing impurities, it competes for natural gas in transport and it is also injected in the gas distribution grids [72].

4.4 Fuel cell applications

A very attractive application of biogas for electricity production is its use in fuel cells [73]. Before gas feed, biogas must be upgraded after the removal of carbon dioxide and sulfur compounds. The classic methane-driven fuel cells convert catalytically methane into a mixture of carbon monoxide and hydrogen. Hydrogen is separated and used as a fuel in a traditional hydrogen/oxygen fuel cell to generate electricity.

The advantages of fuel cell applications with methane consist in their higher efficiency compared to combustion and co-generation [74]. Next, the released heat can be utilized for maintaining the temperature regime in the fuel cell. A disadvantage of this method is the necessity of carbon monoxide removal and the subsequent charging of the atmosphere by carbon dioxide. More attractive is to convert methane (or biogas) into electricity in one step [74–76] in solid oxide fuel cells. However, the power density is still low for practical use.

Besides these applications biogas can be used also for chemical production using dry reforming to produce synthesis gas (a mixture of carbon monoxide and hydrogen) which is further used for light hydrocarbon production by the Fischer-Tropsch process [77].

4.5 Biogas upgrading

Biogas upgrading means the removal of carbon dioxide, partially or completely, and the traces of sulfur compounds. The produced gas is reach of methane and it is competitive to the natural gas in the gas distribution grid. It is suitable for domestic purposes and for transport as well.

Once separated, the resulting methane can be mixed at a desired ration of produce other chemicals via synthesis gas.

The most direct method for biogas upgrading is the membrane separation [78–80]. There is information about commercial equipment for biogas upgrading. Some of them is based on membrane separation [81], or by pressure-swing adsorption to reach capacity from about 500 to 5000 Nm^3/h methane [82].

There are also proposals to recycle CO_2 into methane using bioelectrochemical systems [83]. Although, it seems attractive, energy input is required with the release of carbon dioxide. That is why the use of the PtG approach as mentioned above based on energy surpluses, or other renewable energies, like solar or wind ones are recommended.

4.6 Feasibility of biogas production and use

The feasibility of a biogas equipment depends on different factors. First, it is the amount and generation rate of feedstock (manure, straw, activated sludge, etc.) and its threat to the environment. Next, it is the need of heat or electricity for the considered location. Then, it could be assumed to use biogas as alternative fuel for transport purposes, to be injected in the gas distribution grid or electricity production by co-generation. Biogas upgrade is required it is supposed for transport purposes or for mixing with natural gas in the grid.

It is apparent that different substrates require different approaches. Heating is beneficial because it used to maintain temperature even at mesophilic process. It is possible to maintain it using the heat from a cogeneration (CHP) system after combustion of biogas.

An innovative method is the Power to Gas method integrating the electricity grid with the gas [72].

After selection of biogas as appropriate option for waste treatment with energy recovery, there is necessity to try to select the best method of application for the present community. There are various factors that can affect the rate and amount of produced, namely.

- Type of digester and its capacity
- Temperature (mesophilic or thermophilic process)
- Retention time with the corresponding digester size
- The necessity of pH maintenance
- The presence of certain chemicals in the substrate.

At times of surplus of variable renewable electricity production, hydrogen may be produced via electrolysis, thus storing energy.

5. Conclusions

The biogas has various applications, starting with the simple combustion. Important applications are electricity production by co-generation, by fuel cell applications, as a fuel for transport purposes and as a feedstock for production of chemicals like light hydrocarbons. Prior to its use as a fuel or for chemical purposes, upgrading of biogas with removal of carbon dioxide is desirable. A promising approach is "power to gas" process after electricity production for recycling of carbon dioxide into methane. The biogas yield and quality depend either by the pre-treatment, or the operation mode (substrate dosage, choice of anaerobic digester, etc.). It seems that simple enzyme pre-treatment is good enough compared with more sophisticated methods, like ultrasonic or microwave treatment, even steam explosion. The choice of methods and scale of application depends on the regional raw material access, the energy demands and climate peculiarities.

Acknowledgements

The author kindly acknowledges for financial support to project № BG05M2OP001-1.002-0014 „Center of competence HITMOBIL - Technologies and systems for generation, storage and consumption of clean energy", funded by Operational Programme "Science and Education For Smart Growth" 2014-2020, co-funded by the EU from European Regional Development Fund.

Author details

Venko Beschkov
Institute of Chemical Engineering, Bulgarian Academy of Sciences, Sofia, Bulgaria

*Address all correspondence to: vbeschkov@iche.bas.bg

References

[1] Yentekakis IV, Grammatiki G. Biogas management: Advanced utilization for production of renewable energy and added-value chemicals. Frontiers in Environmental Science. 2017;**5**:7

[2] Van Hook JP. Methane-steam reforming. Catalysis Reviews: Science and Engineering. 1980;**21**:1-51

[3] Zhang Y, Wang W, Wang Z, Zhou X, Wang Z, Liu CJ. Steam reforming of methane over Ni/SiO2 catalyst with enhanced coke resistance at low steam to methane ratio. Catalysis Today. 2015; **256**:130-136

[4] de Jong E, Jungmeier G. Biorefinery concepts in comparison to petrochemical refineries, Chapter 1. In: Industrial Biorefineries and White Biotechnology. Amsterdam: Elsevier B. V. 2015. pp. 3-33. DOI: 10.1016/B978-0-444-63453-5.00001

[5] Alemán-Nava GS, Meneses-Jácome A, Cárdenas-Chávez DL, Díaz-Chavez R, Scarlat N, Dallemand JF, et al. Bioenergy in Mexico: Status and perspective. Biofuels, Bioprod. Bioref. 2015;**9**:8–20. DOI: 10.1002/bbb.1523.

[6] Virmond E, Rocha JD, Moreira RFPM, José HJ. Valorization of agroindustrial solid residues and residues from biofuel production chains by thermochemical conversion: A review, citing Brazil as a case study. Brazilian Journal of Chemical Engineering. 2013;**30**:197-229

[7] Aresta M. Anaerobic fermentation: Biogas from waste – the basic science. Chapter 17. In: Aresta M, Dibenedetto A, Dumeignil F, editors. Biorefinery: From Biomass to Chemicals and Fuels. Berlin: W. De Gruyter GmbH, 2012. pp. 377-403

[8] Tabatabaei M, Aghbashlo M, Valijanian E, Kazemi H, Panahi S, Nizami AS, et al. A comprehensive review on recent biological innovations to improve biogas production, Part 1. Upstream strategies. Renewable Energy. 2020;**146**:1204-1220

[9] Indran S, Divya D, Rangappa SM, Siengchin S, Merlin Christy SP, Gopinath LR. Perspectives of anaerobic decomposition of biomass for sustainable biogas production: A review. E3S Web Conferences. E3S Web Conferences, Volume 302, 2021. DOI: 10.1051/e3sconf/20213020101

[10] Vasco-Correa J, Khanal S, Manandhar A, Shah A. Anaerobic digestion for bioenergy production: Global status, environmental and techno-economic implications, and government policies. Bioresource Technology. 2018;**247**:1015-1026

[11] Appels L, Lauwers J, Degrève J, Helsen L, Lievens B, Willems K, et al. Anaerobic digestion in global bio-energy production: Potential and research challenges. Renewable and Sustainable Energy Reviews. 2011;**15**:4295-4301

[12] Holm-Nielsen JB, Al Seadi T, Oleskowicz-Popiel P. The future of anaerobic digestion and biogas utilization. Bioresource Technology. 2009;**100**:5478-5484. DOI: 10.1016/j.biortech.2008.12.046

[13] Slusarz G, Gołebiewska B, Cierpiał-Wolan M, Gołebiewski J, Twaróg D, Wójcik S. Regional diversification of potential, production and efficiency of use of biogas and biomass in Poland. Energies. 2021;**14**:742. DOI: 10.3390/en14030742

[14] Prajapati KK, Pareek N, Vivekanand V. Pretreatment and multi-feed anaerobic co-digestion of agro-industrial residual biomass for improved biomethanation and kinetic analysis. Frontiers in Energy Research. 23 October 2018.

[15] Angelidaki I, Ellegaard L, Ahring BK. Applications of the anaerobic digestion process. In: Ahring BK, editor. Biomethanation II. Advances in Biochemical Engineering/ Biotechnology. Vol. 82. Berlin, Heidelberg: Springer; 2003. DOI: 10.1007/3-540-45838-7_1

[16] Hashimoto AG. Methane from cattle waste: Effects of temperature, hydraulic retention time, and influent substrate concentration on kinetic, parameter (K). Biotechnology and Bioengineering. 1982; **24**:2039-2052

[17] Baader W, Done E, Brenndoerfer M. Biogas in Theorie und Praxis, Behandlung organ. Reststoffe aus d. Landwirtschaft durch Methangärung, KTBL-Schrift. Mǜnster: Landwirtschaftsverlag 229. 2021

[18] Cseh T, Czako L, Toth J, Tengerdy RP. Two-phase anaerobic fermentation of liquid swine waste to methane. Biotechnology and Bioengineering. 1984;**26**:1425-1429

[19] Hills DJ. Environment and health. Chemical characteristic of and methane production from turkey manure. Poultry Science. 1982;**61**:677-682

[20] Nwokolo N, Mukumba P, Obileke KC, Enebe M. Waste to energy: A focus on the impact of substrate type in biogas production. PRO. 2020;**8**:1224. DOI: 10.3390/pr8101224

[21] Garcia-Heras JL. Reactor sizing, process kinetics, and modelling of anaerobic digestion of complex wastes. In: Mata-Alvarez J, editor. Biomethanization of the Organic Fractions in Municipal Solid Wastes. London, U.K: IWA-Publishing; 2003

[22] Krishania M, Kumar V, Kumar Vijay V, Malik A. Analysis of different techniques used for improvement of biomethanation process: A review. Fuel. 2013;**106**:1-9

[23] Heo NH, Park SC, Lee JS, Kang H. Solubilization of waste activated sludge by alkaline pretreatment and biochemical methane potential (BMP) tests for anaerobic co-digestion of municipal organic waste. Water Science and Technology. 2003;**48**:211-219

[24] Li Z, Li L, Zheng M, Fu G, Lar JS. Anaerobic co-digestion of cattle manure with corn stover pretreated by sodium hydroxide for efficient biogas production. Energy & Fuels. 2009;**23**: 4635-4639

[25] Baccay RA, Hashimoto AG. Acidogenic and methanogenic fermentation of causticized straw. Biotechnology and Bioengineering. 1984;**26**:885

[26] Wu S-L, Wei W, Ni B-J. Enhanced methane production from anaerobic digestion of waste activated sludge through preliminary pretreatment using calcium hypochlorite. Journal of Environmental Management. 2021;**295**: 113346

[27] Matsushita Y, Inomata T, Hasegawa T, Fukushima K. Solubilization and functionalization of sulfuric acid lignin generated during bioethanol production from woody biomass. Bioresource Technology. 2009; **100**:1024-1026

[28] Zhu S, Wu Y, Ziniu Y, Liao J, Zhang Y. Pretreatment by microwave/ alkali of rice straw and its enzymic hydrolysis. Process Biochemistry. 2005; **40**:3082-3086

[29] García-Cubero MT, González-Benito G, Indacoechea I, Coca M, Bolado S. Effect of ozonolysis pretreatment on enzymatic digestibility of wheat and rye straw. Bioresource Technology. 2009;**100**:1608-1613

[30] Lee JM, Jameel H, Venditti RA. Effect of ozone and autohydrolysis pretreatments on enzymatic

digestibility of coastal Bermuda grass. BioResources. 2010;**5**:1084-1101

[31] Dadi A, Schall C, Varanasi S. Mitigation of cellulose recalcitrance to enzymatic hydrolysis by ionic liquid pretreatment. Applied Biochemistry and Biotechnology. 2007;**137-140**: 407-421

[32] Heinze T, Schwika K, Barthel S. Ionic liquids as reaction medium in cellulose functionalization. Macromolecular Bioscience. 2005;**5**: 520-525

[33] Li Q, He YC, Xian M, Jun G, Xu X, Yang JM. Improving enzymatic hydrolysis of wheat straw using ionic liquid 1-ethyl-3-methyl imidazolium diethyl phosphate pretreatment. Bioresource Technology. 2009;**100**: 3570-3575

[34] Rogalinski T, Ingram T, Brunner GJ. Hydrolysis of lignocellulosic biomass in water under elevated temperatures and pressures. Journal of Supercritical Fluids. 2008;**47**:54-63

[35] van Walsum G, Allen S, Spencer M, Laser M, Antal M, Lynd L. Conversion of lignocellulosics pretreated with liquid hot water to ethanol. Applied Biochemistry and Biotechnology. 1996, 1996;**57-58**:157-170

[36] Das AC, Mondal C, Roy S. Pretreatment methods of ligno - cellulosic biomass: A review. Journal of Engineering Science and Technology. 2015;**8**:141-165

[37] Sun Y, Cheng J. Hydrolysis of lignocellulosic materials for ethanol production: A review. Bioresource Technology. 2002;**83**:1-11

[38] Grous WR, Converse AO, Grethelin HE. Effect of steam explosion pretreatment on pore size and enzymatic hydrolysis of poplar. Enzyme and Microbial Technology. 1986;**8**:274-280

[39] Morjanoff PJ, Gray PP. Optimization of steam explosion as method for increasing susceptibility of sugarcane bagasse to enzymatic saccharification. Biotechnology and Bioengineering. 1987;**29**:733-741

[40] Mackie KL, Brownell HH, West KL, Saddler JN. Effect of sulphur dioxide and sulphuric acid on steam explosion of aspenwood. Journal of Wood Chemistry and Technology. 1985;**5**:405-425

[41] McMillan JD. Pretreatment of lignocellulosic biomass. In: Himmel ME, Bake JO, Overend RP, editors. Enzymatic Conversion of Biomass for Fuels Production. Washington, DC: American Chemical Society; 1994. pp. 292-324

[42] Timilsena YP. Effect of different pretreatment methods in combination with the organosolv delignification process and enzymatic hydrolysability of three feedstocks in correlation with lignin structure [thesis]. Thailand: Asian Institute of Technology; 2012

[43] Zheng Y, Pan Z, Zhang R. Overview of biomass pretreatment for cellulosic ethanol production. International Journal of Agricultural Biological Engineering. 2009;**2**:51-68

[44] Lee J-W, Gwak K-S, Park J-Y, Park M-J, Choi D-H, Kwon M, et al. Biological pretreatment of softwood Pinus densiflora by three white rot fungi. Journal of Microbiology. 2007;**45**: 485-491

[45] Yang CP, Shen ZQ, Yu GC, Wang J. Effect and aftereffect of γ-radiation pretreatment on enzymatic hydrolysis of wheat straw. Bioresource Technology. 2008;**99**:6240-6245

[46] Nitayavardhana S, Rakshit SK, Grewell D, van Leeuwen JH, Khanal SK. Ultrasound pretreatment of cassava chip slurry to enhance sugar release for subsequent ethanol production.

Biotechnology and Bioengineering. 2008;**101**:487-496

[47] Kumar P, Barrett DM, Delwiche MJ, Stroeve P. Pulsed electric field pretreatment of switchgrass and woodchips species for biofuels production. Industrial & Engineering Chemistry Research. 2011;**50**: 10996-11001

[48] Binod P, Satyanagalakshmi K, Sindhu R, Janu KU, Sukumaran RK, Pandey A. Short duration microwave assisted pretreatment enhances the enzymatic saccharification and fermentable sugar yield from sugarcane bagasse. Renewable Energy. 2012;**37**: 109-116

[49] Ha SH, Mai NL, An G, Koo YM. Microwave-assisted pretreatment of cellulose in ionic liquid for accelerated enzymatic hydrolysis. Bioresource Technology. 2011;**102**: 1214-1219

[50] Xu J, Chen H, Kadar Z, Thomsen AB, Schmidt JE, Peng H. Optimization of microwave pretreatment on wheat straw for ethanol production. Biomass and Bioenergy. 2011;**35**:3859-3864

[51] Angelov I, Beschkov V. Biogas production by plant waste treatment. Dubai, UAE: 11th International Conference on Biofuel & Bioenergy; 19-20 February, 2020

[52] Manimegalai R, Gopinath LR, Merlin Christy P, Divya D. Isolation and identification of acetogenic and methanogenic bacteria from anoxic black sediments and their role in biogas production. International Journal of Plant, Animal & Environmental Sciences. 2014;**4**:156-164

[53] Siegert I, Banks C. The effect of volatile fatty acid additions on the anaerobic digestion of cellulose and glucose in batch reactors. Process Biochemistry. 2005;**40**:3412-3418

[54] Fuchs G, Thauer R, Ziegler H, Stickler N. Carbon isotope fractionation in Methanobacterium thermoautotrophicum. Archives of Microbiology. 1979;**120**:135-137

[55] Conrad R. Quantification of methanogenic pathways using stable carbon isotopic signatures: A review and a proposal. Organic Geochemistry. 2005;**36**:739-752

[56] Seghezzo L, Zeeman G, van Lier JB, Hamelers HVM, Lettinga G. A review: The anaerobic treatment of sewage in UASB and EGSB reactors. Bioresource Technology. 1998;**65**:175-190

[57] Grobicki A, Stuckey DC. Hydrodynamic characteristics of the anaerobic baffled reactor. Water Research. 1992;**26**:371-378

[58] Beschkov V. In: Jacob-Lopes E, Zepka LQ, editors. Biogas, Biodiesel and Bioethanol as Multifunctional Renewable Fuels and Raw Materials. Rijeka, Croatia: InTech; 2017. pp. 185-205. Chapter 10

[59] Beschkov V, Sapundzhiev T, Angelov I. Modelling of biogas production from glycerol by anaerobic process in a baffled multi-stage digestor. Biotechnology and Biotechnological Equipment. 2012;**26**:3244-3248

[60] Beschkov V, Angelov I, Petrova P. Biogas production from glycerol in a multistage anaerobic digestor. Current Topics in Biotechnology. 2012;**7**:61-69

[61] Ashby RD, Wyatt VT, Foglia TA, Solaiman DKY. Industrial products from biodiesel glycerol. In: Hou CT, Shaw JJ, editors. Biocatalysis and Bioenergy. Vol. 2009. Hoboken: J. Wiley & Sons Inc; 2009. pp. 131-154

[62] da Silva GP, Mack M, Contiero J. Glycerol: A promising and abundant carbon source for industrial microbiology. Biotechnology Advances. 2009;**27**:30-39

[63] Biebl H, Menzel K, Zeng AP, Deckwer WD. Microbial production of 1,3-propanediol. Applied Microbiology and Biotechnology. 1999;**52**:289-297

[64] Kolesárová N, Hutňan M, Špalková V, Lazor M. Biodiesel by-Products as Potential Substrates for Biogas Production. Tatranske´ Matliare Slovakia: Proceedings of the 37th International Conference of SSCHE, 14–28 May; 2010. pp. 1126-1139

[65] López JAS, de los Ángeles Martín Santos M, AFC P, Martín AM. Anaerobic digestion of glycerol derived from biodiesel manufacturing. Bioresource Technology. 2009;**100**: 5609-5615

[66] Adiyia PK. Anaerobic co-digestion of digestate with glycerol to enhance biogas production [thesis]. Sweden: Linköpings Universitet; 2021

[67] Sittijunda S, Sitthikitpanya N, Plangklang P, Reungsang A. Two-stage anaerobic codigestion of crude glycerol and micro-algal biomass for biohydrogen and methane production by anaerobic sludge consortium. Fermentation. 2021;7:175. DOI: 10.3390/fermentation7030175

[68] Wohlgemut O. Co-digestion of hog manure with glycerol to boost biogas and methane production [thesis]. Canada: University of Manitoba; 2009

[69] Fountoulakis MS, Manios T. Enhanced methane and hydrogen from municipal solid waste and agro-industrial by-products co-digested with crude glycerol. Bioresource Technology. 2009, 2009;**100**:3043-3047

[70] de Souza RN, Mambeli Barros R, Silva dos Santos IF, GLT F, da Silva SPG. Electric energy generation from biogas derived from municipal solid waste using two systems: Landfills and anaerobic digesters in the states of Sao Paulo and Minas Gerais, Brazil. Sustainable Energy Technologies and Assessments. 2021;**48**:101552

[71] Biogas Cogeneration [Internet] 2021. Available from: https://www.biogas-renewable-energy.info/biogas_cogeneration.html [Accessed October 10, 2021].

[72] Liebetrau J, Kornatz P, Baier U, Wall D, Murphy JD. In: Murphy JD, editor. Integration of Biogas Systems into the Energy System: Technical aspects of flexible plant operation. Vol. 37. IEA Bioenergy Task; 2020. p. 8

[73] O'Hayre RP, Cha S-W, Colella WG, Prinz FB. Fuel Cell Fundamentals. 2nd ed. N. Jersey, USA: J. Wiley & Sons, Inc; 2009. pp. 384-385

[74] Perry Murray E, Tsai T, Barnett SA. A direct-methane fuel cell with a ceria-based anode. Nature. 1999;**400**:649-651

[75] Yu C, deGlee B, Yu T, Wang Z, Zhao B, Wei Y, et al. A robust fuel cell operated on nearly dry methane at 500°C enabled by synergistic thermal catalysis and electrocatalysis. Nature Energy. 2018;**3**:1042-1050

[76] Putna ES, Stubenrauch J, Vohs JM, Gorte RJ. Ceria-based anodes for the direct oxidation of methane in solid oxide fuel cells. Langmuir. 1995;**11**: 1832-1837

[77] Cassinelli WH, Damyanova S, Parizotto NV, Zanchet D, Bueno JMC, Marques CMP. Study of the properties of supported Pd catalysts for steam and autothermal reforming of methane. Applied Catalysis A: General. 2014;**475**: 256-2695

[78] Chen XY, Vinh-Thang H, Ramirez AA, Rodrigue D, Kaliaguine S. Membrane gas separation technologies for biogas upgrading. RSC Advances. 2015;**5**:24399-24448

[79] Makaruk A, Miltner M, Harasek M. Membrane biogas upgrading processes for the production of natural gas. Substitute, Separation & Purification Technology. 2010;**74**:83-92

[80] Garcia Gomes M, de Morais LC, Pasquini D. Use of membranas for biogas purification: Review. Holos Environment. 2019;**19**:466-501. DOI: 10.14295/holos.v19i3.12339

[81] Bright Biomethane. Enschede [Internet]: 2021. Available from: https://www.brightbiomethane.com/technology-biogas-upgrading/). [Accessed: October 5, 2021]

[82] PSA Biogas Upgrading and Purification Plants. Stuttgart: Mahler AGS. Available from: https://www.mahler-ags.com/biogas/psa; [Accessed: October 05, 2021]

[83] Liu D, Zeppilli M, Villano M, Buisman C, ter Heijne A. Methane production at biocathodes: Principles and applications, Chapter 5. In: Wang A, Liu W, Zhang B, Cai W, editors. Bioelectrosynthesis: Principles and Technologies for Value-Added Products. Weinheim: Wiley-VCH Verlag, GmbH & Co; 2020

Chapter 3

Resource Reclamation for Biogas and Other Energy Resources from Household and Agricultural Wastes

Donald Kukwa, Maggie Chetty, Zikhona Tshemese, Denzil Estrice and Ndumiso Duma

Abstract

The chapter's goal is to highlight how the reclamation of household and agricultural wastes can be used to generate biogas, biochar, and other energy resources. Leftover food, tainted food and vegetables, kitchen greywater, worn-out clothes, textiles and paper are all targets for household waste in this area. Agricultural waste includes both annual and perennial crops. Annual crops are those that complete their life cycle in a year or less and are comparable to bi-annual crops, although bi-annuals can live for up to two years before dying. The majority of vegetable crops are annuals, which can be harvested within two to three months of seeding. Perennials crops are known to last two or more seasons. Wastes from these sources are revalued in various shapes and forms, with the Green Engineering template being used to infuse cost-effectiveness into the process to entice investors. The economic impact of resource reclamation is used to determine the process's feasibility, while the life cycle analysis looks at the process's long-term viability. This is in line with the United Nations' Sustainable Development Goals (SDGs), whose roadmap was created to manage access to and transition to clean renewable energy by 2030, with a target of net zero emissions by 2050.

Keywords: food waste, clothes and textiles, annual and perennial crops, post-harvest waste, green engineering, biogas and biochar, economic impact, life cycle analysis

1. Introduction

The population growth rate is an important factor to consider when examining the past, present, and future resource base for sustainability. The increase in global population, combined with increased agricultural productivity and medical advancements, has resulted in resource consumption exceeding the environment's carrying capacity [1]. As the human population expands, so does the potential for tremendous, irreversible changes. Increased biodiversity loss, greenhouse gas emissions, worldwide deforestation, stratospheric ozone depletion, acid rain,

IntechOpen

topsoil loss, and water, food, and forest resource shortages are all signs of severe environmental stress in many parts of the world [2]. This human impact on the environment informs the current biotic and abiotic resource depletion.

Biotic resources are resources that come from the biosphere, which are living or once-living beings and forests, as well as the materials that come from them in the ecosystem [3]. Biotic resources include forests and forest products, crops, birds, wildlife, fish, and other marine life. These resources rejuvenate and duplicate themselves, making them renewable. Fossil fuels like coal, natural gas, petroleum, etc. are biotic resources as well, but they are non-renewable; as non-renewable resources get depleted, human society will increasingly rely on the self-renewing capacity of biotic resources [4]. Abiotic resources, on the other hand, are usually obtained from the lithosphere, atmosphere, and hydrosphere. Examples of abiotic resources are water, air, soil, sunlight, radiation, temperature, atmosphere, humidity, acidity and mineral raw materials [5].

Household waste is defined by Reddy [6] and Viljoen et al. [7] as waste generated by household activities such as cooking, sweeping, cleaning, fuel burning, repairs, and gardening. Old clothing, old furnishings, retired equipment, glass, paper, metal packaging, and old books and newspapers are all examples of used products or materials.

Over the last few years, the reuse of home garbage, harvest, post-harvest, and forest leftovers has gained popularity. This is to close the energy gap that has been formed as a result of rising demand from the rural-urban migration and the general improvement in the human population's lifestyle. The demand for high-quality, high-value items has put a strain on scientific research and the manufacturing sector of the economy, threatening to deplete fossil resources. Resource reclamation for benefit employs labour and generates income [8].

The strategic approach of the National Waste Management Strategy (NWMS) of South Africa 2020 to waste management adopted the circular economy approach, in which there is no waste. In a circular economy, any material whose value has degraded for a given application becomes a raw material for another process [7]. This chapter highlights the reclamation of waste resources from agricultural harvest and post-harvest operations for energy resources such as biogas and biohydrogen. Also, the flue gas from the production of biochar was harnessed. Household wastes were also harnessed for biogas generation. The chapter also streamlined the economic benefits of waste resources reclamation and the advantages of a circular economy. The life cycle analysis looked at the composition of household and farm wastes, and the volumetric flow characteristics of the waste materials.

2. The characteristics and types of wastes

2.1 The characteristics of wastes

Wastes are often characterized by professionals depending on the sources from which they are created. Household rubbish, hospitals, agricultural waste, industrial waste, mining activities, public spaces, and other sources all contribute to waste production. Wastes are toxic by nature and can harm the environment as well as animal health.

2.1.1 Household waste

Household waste is any waste that is generated from running a domestic facility, accounting for more than two-thirds of the municipal solid waste (MSW) stream [9].

It can include food materials, plastics, cardboard, rubber, metal, paper, wood, fabric, chemicals etc. Hazardous substances in household waste, unlike waste streams from industrial sources, are not strictly regulated under hazardous waste regulations. As a result, household hazardous waste (HHW) is dumped in landfills alongside general household waste (HW). Cleaning products, self-care products, pharmaceuticals, home-care products, automotive maintenance products, electronic equipment, and general maintenance products for machinery are examples of items that are frequently used in places of residence, commercial centres, corporate organizations, and institutions. These products contain substances that, on their own or when combined with others, produce secondary compounds capable of causing severe environmental and public health damage [10].

2.1.2 Healthcare waste

Surgical trash, blood, body parts, medications, wound dressing materials, syringes, and needles are all examples of hospital waste. Hospitals, clinics, veterinary hospitals, and medical laboratories all produce this form of trash. Contamination and illness are common outcomes of hospital waste [4].

2.1.3 Agricultural waste

Agricultural waste is produced by farming, animal husbandry, and market gardens, among other activities. Pesticide containers, expired medications and wormers, extra milk, corn husks, corn cubs, corn silage, rice husks, rice straw, and other agricultural wastes are the most prevalent [1].

2.1.4 Industrial waste

Industries generate a wide range of trash. Petroleum refineries, chemical plants, cement factories, power plants, textile mills, and food processing and beverage facilities are all industrial waste generators. These industries produce a considerable amount of waste, which impairs the environment's esthetics and may have an influence on the chemistry of the atmosphere [1, 2].

2.1.5 Commercial waste

The volume of items purchased and sold, as well as technical improvements in industry and transportation, all contribute to commercial waste [4]. Food, textiles, discarded household and medical supplies, and a range of other objects might be included.

2.1.6 Electronic waste

Discarded old electronic equipment such as televisions, microwaves, vacuum cleaners, and music players are examples of electronic waste sources. E-scrap, or waste electrical and equipment, is another name for it. These wastes are high in cadmium, lead, and mercury, all of which are toxic to persons and the environment [7].

2.1.7 Mining and quarrying wastes

Mine wastes are coarse wastes generated during the mining stages of rock blasting and tunnel preparation, as well as tailings from ore processing. Overburden materials that must be removed and disposed of to get access to ore or precious rock are known as quarrying wastes [10]. Two examples are toxic gases created during blasting and other mining contaminants. The impact of mining waste on the local environment and surroundings is significant.

2.1.8 Demolition and construction wastes

Depending on the project, bricks and masonry, concrete, wood, metal (including plumbing), plaster and drywall, glass and windows, demolition debris, and other demolition and building materials may be employed. Garbage like asphalt, rubble, tile., etc., from huge projects, as well as construction and building materials trash such as packing boxes, concrete debris, plastics, and wood [8].

2.1.9 Radioactive waste

Gamma rays, alpha particles, beta particles, and neutron radiation are all forms of radiation produced by radioactive waste. Radioactive waste is produced by nuclear reactors or atomic explosions, and it is particularly harmful to animals. High-level waste, low-level waste, and transuranic waste are the three categories of radioactive waste. This technology is used in the power generating sector of the economy as well as the radiological unit in hospitals for imaging diagnosis [10, 11].

2.2 Types of waste

Professionals also have characterized waste according to (i) the physical states of materials namely solid, liquid and gas; and (ii) the potential of microbial attack namely biodegradable and nonbiodegradable materials.

2.2.1 Solid wastes

Solid wastes account for the majority of the trash produced by human civilisation. Agricultural wastes, domestic wastes, radioactive wastes, industrial wastes, and biomedical wastes are examples of solid wastes that can be categorized based on their source or nature [1, 4].

2.2.2 Liquid wastes

Liquid wastes are wastes that are formed in a liquid state as a result of industrial production, washing, flushing, or other industrial activities. Liquid waste is also produced in significant amounts by households. Used vegetable oil and kitchen wastewater are two examples [1].

2.2.3 Gaseous wastes

The principal sources of gaseous wastes include internal combustion engines, incinerators, coal-fired power plants, and industrial processes. Depending on their qualities, gaseous wastes might be odiferous or toxic. Smog and acid precipitation develop when they mix with other gases [10].

2.2.4 Biodegradable wastes

These are wastes that originate from the kitchen, such as food scraps, garden trash, and so on. Moist trash, green waste, recyclable waste, food waste, and organic waste are all terms used to describe biodegradable garbage. This may be composted to produce manure, also known as humus. Biodegradable wastes break down over time, depending on the substance and can be destroyed by biotic and abiotic factors such as microorganisms (e.g. bacteria, fungus), temperature, ultraviolet radiation, oxygen, and others. They are digested anaerobically to provide energy in the form of heat, electricity, and fuel [1, 6, 11].

2.2.5 Non-biodegradable wastes

Non-biodegradable wastes are those that are not easily degraded by natural agents or dissolved by them. They stay undamaged for many years and are the primary sources of pollution in the air, water, and soil, as well as illnesses such as cancer [6, 11]. Dry waste refers to non-biodegradable waste. Newspapers, shattered glass shards, and plastics, which are employed in practically every sector, are all good examples. Cans, metals, and agricultural and industrial chemicals are further examples. Dry wastes are recyclable and reusable. Non-biodegradable trash is bad for the environment, thus there's a rising demand for alternatives. In response, biodegradable polymers (also known as biocomposites) have evolved, although they remain prohibitively costly [12]. Polymers are the backbones of plastic materials, and they are used in an ever-growing number of applications.

3. Household and agricultural waste resources

Household garbage has become one of the most prominent sources of serious impairment to the rural environment due to huge amounts of rubbish discharged and improper disposal. The amount of rubbish created rises in lockstep with the world's population. Household garbage production will have grown by about 70% per year by 2050, suggesting that waste production will have surpassed population growth by more than twice [1, 4, 7]. Household trash management is a tough task due to the rising volume of rubbish produced throughout the world and the vast variety of different components included in this waste stream. Sorting rubbish at the source is crucial for recycling and the circular economy to thrive [9]. Agricultural waste consists crop remnants, weeds, leaf litter, sawdust, forest detritus, and animal manure. Waste from agro-based industries such as palm oil, rubber, and wood processing factories has increased by several times as a result of increased agric mechanization and automation. Significant quantities of phosphate and nitrogen, as well as biodegradable organic carbon, pesticide residues, and fecal coliform bacteria, are found in agricultural wastes that run straight into surface waters [6].

3.1 Household waste resources

Household wastes can be solid, and liquid. The different categories of household waste are addressed in the following sub-sections:

3.1.1 Household solid waste resources

Many cities in developing countries have a challenge of improper management of solid household waste which is a constituent of municipal solid waste [11, 12]. Improper management is because there is usually a lack of understanding of the waste generation and its composition which leads to municipal authorities being unable to establish and execute efficient management plans [13]. This lack of understanding means that authorities most often use equipment and management plans that are not tailored for some communities/cities [14, 15]. Solid household waste is a broad term for solid waste materials found in a home which can be characterized into different classes such as organics, plastic, paper, glass and ceramics, metals and tins, and other types of wastes. Some of these categories can be sub-divided into more specific products such as food waste, garden waste, magazine, newspaper, office paper, miscellaneous paper all being organic waste [16]. Plastic waste includes bottles, containers, jars and bags while paper waste contains cardboard, packaging material, newspapers. Other waste material includes disposable diapers and sanitation waste [17].

Solid household waste composition and quantities produced are influenced by socio-economic dynamics such as family size, income, car ownership, age, education etc. [18]. This evidence has been shown in studies where overall waste generation and generation of individual components of waste streams have differed between the less and more prosperous sectors of a city [19, 20]. Research has also shown that lower-middle-class communities generate waste with a high potential of recyclability [21]. Solid household waste has been identified as a huge contributor (82%) of the total solid waste compared to waste from commercial, institutions and industrial locations [22]. Different strategies have been developed for resolving the challenges of waste including solid household waste. It is well known for example that plastic and its related materials, glass and ceramics are non-degradable, however can be recycled into new products instead of being thrown into dumping sites as they have incredibly negative impacts on the environment [23, 24].

The organic part of solid household waste (about 68%) is biodegradable and therefore presents a great opportunity to be further used as a resource. This organic-rich waste is a good medium for microbial growth, consequently, it can be used to produce energy (in the form of biogas) which is an excellent provision for positive contribution to the environmental, energy and economic needs [25, 26]. Energy derived from household waste becomes very significant since sufficient energy access is one of the crucial factors of improvement in any country in the world. In this way of economic development, the fight against poverty, education and adequate healthcare is facilitated [27]. Biogas is a result of a four-step (hydrolysis, acidogenesis, acetogenesis and methanogenesis) microbially aided anaerobic digestion process with approximately 50–70% methane, 30–45% carbon dioxide and some trivial amount of other trace elements [28, 29]. Biogas is produced from organic substrates and therefore the resulting waste is rich in nutrients that are used as bio fertilizer [30].

Liu et al. [31] have produced hydrogen and methane from solid household waste using a two-stage fermentation process. In another study, solid household waste consisting of 80.4% organic matter has been used to produce biogas through an anaerobic batch reactor [32].

3.1.2 Household liquid waste resources

Liquid products are used in common rooms of a household such as a kitchen, bathroom, garages as well as basements. These products have the potential of causing serious environmental and health problems both during their time of use as well as after they have been discarded [33]. Often the consumer of these products is not aware of how to properly dispose of them after usage. The need for identifying appropriate ways of discarding liquid household waste has been realized when serious health problems and damage to areas of disposal started to manifest [34].

Liquid household waste incorporates any liquid waste from places such as the kitchen (cooking oil, dish detergents, floor cleaning products, microwave/oven cleaners, furniture and metal polishes, drain cleaners, etc.), bathroom (health and beauty products, disinfectants, basin and tub detergents and toilet bowl cleaners), laundry room (bleaches, fabric softeners, detergents, spot removers, etc.) as well as the garages and/basements (paints, pesticides and herbicides, lawn and garden care products, fuel, oils, glues and adhesives, etc.) [34, 35]. Most of these are made from hazardous chemicals although they can be paid for over the counter by any person from supermarkets, automotive centres and hardware stores.

The negative impacts to surface water, groundwater and the soil caused by improper disposal of liquid household waste have brought about the need to look for solutions to the challenge of waste management [36]. Consequently, research has been done across the globe and solutions are slowly being realized and embraced. These include strict prevention, reduction at source, treatment of liquid waste before disposal, recycling the waste into other useful products, valorisation of waste as a form of meeting other demands (energy demands) in societies [37, 38]. For example, liquid waste is used to produce biogas, electrical energy and heat through the following processes; anaerobic fermentation, pyrolysis, biothermal composting, hydrothermal destruction,etc. [39, 40]. The realization that household liquid waste is a renewable energy source is the beginning of solving socio-economic issues because the whole technology employs people in the production of both the technology gear as well as energy thus addressing environmental issues while benefiting the economy [41].

Kitchen wastewater has been used as a substrate for the production of biogas by Kumar et al. [42] where the Up Flow—Anaerobic Sludge Blanket (USAB) reactor was employed. Another study that employed kitchen wastewater co-digested it with several other substrates such as water hyacinth, cow manure and sewage sludge for biogas production which had 60–65% methane, 14–18% carbon dioxide as well as 20–21% other gases [43]. Domestic liquid waste has been used as a constituent of the municipal liquid waste for the production of electricity in sufficient volumes to lessen the electrical load of the water treatment plant while producing surplus power to feed into the grid [44].

3.2 Agricultural waste resources

Modern agriculture depends primarily on annual crops, which are crops that can be harvested within two to three months of seeding. Annual crops live their whole life cycle in a year or less. These crops are typically classified as summer crops (warm-season) and winter crops (cool-season crops). Warm-season crops develop faster during warmer times of the year and are typically seed and fruit crops. Cool-season crops develop faster during cooler times of the year and are typically root, leaf, flower bud, and stem crops. Examples of annual crops include onions, tomatoes, popcorn, carrots, peas, kale, and corn [45]. These crops have a lower water requirement and tend to generate more crops produced per drop of water [46]. Bi-annual crops are comparable to annual crops, but they can live up to two years before dying [45]. The first year of bi-annuals results in a short stem and leaves which eventually bloom in the second year.

A more sustainable alternative to annual crops that have been advocated for is perennial crops. Examples of perennial crops include sugarcane [47], coconuts, pineapple, peppermint, spearmint [48], apples, and peaches or apricot [49]. These crops last two or more seasons and are planted once and harvested every year. They lower the chances of soil erosion and limit losses of water and nutrient due to the greater root mass nature of perennials. Perennial crops are preferred for both the quality of the product harvested and their total production [49]. Shifting to perennial crops may enhance many ecosystems services but this will come at a cost as perennial crops have higher water requirements. Furthermore, lower yields that are more stable than those of annual crops can be expected [46].

3.2.1 Farm produce waste

Farm waste is classified under the agricultural waste stream. Agricultural waste refers to the residues produced from growing and processing crops. They are the non-products of production [50]. Agricultural waste includes natural (biodegradable) and non-natural wastes (inorganic) which are produced from agricultural activities such as horticulture, dairy farming, livestock breeding, seed germination, nursery plots, market gardens, grazing land, and forestry. This waste comes as either liquids, solids, and slurries, or sludge. Agricultural waste can be classified according to the activity undertaken [51]. For example, crop production and harvest, sugar processing, fruit and vegetable processing, animal production, rice production, dairy product processing, and coconut production. Each of these activities generates its unique wastes.

The global agricultural waste production has been estimated to be approximately 998 million tonnes yearly. This estimate is likely to increase if farming systems are intensified in developing countries. The total solid wastes produced in any farm includes up to 80% of organic waste and the generation of manure can amount to 5.27/kg/day/1000 kg live weight, on a wet weight basis [50]. Agricultural waste tends to pose serious problems to the environment and humans due to it being toxic, especially the waste that includes pesticides, insecticides, etc. It also has a high pollution potential over extended periods, threatening surface water, underground water, and soil resources [51].

Agricultural waste can be used in various applications. These include fertilizer application (employing animal manures), anaerobic digestion (generating methane gas from manures), pyrolysis (generating bio-oil, char, and gas), animal feed, adsorbents to eliminate heavy metals (used in the adsorption process), and direct combustion [50]. The utilization of waste must either happen rapidly or the waste must be stored under controlled conditions to avoid spoilage of the residues.

3.2.2 Post-harvest wastes

In a world that is ever-growing in population which results in an increasing demand for food, postharvest waste is a critical phenomenon worth addressing. At the forefront of the bio-economy sector are plans to minimize the quantity of waste produced, advance the inescapable waste produced as a resource, and attain noteworthy levels of safe disposal and recycling [52]. Postharvest waste is defined as the amount of food wasted or lost throughout the food chain, after harvesting till consumption. Roughly a third of the food produced on a global scale for human consumption gets wasted or lost, which is about 1.3 billion tonnes of the harvest lost annually [53, 54]. Postharvest waste is intentional. Generated products can be rejected and discarded by growers, distributors, retailers, and consumers if they fail to meet established preferences [55].

Post-harvest wastes are a major contributor to agricultural biomass loss. This is true for the whole world but postharvest losses and waste are more prevalent in developing countries because of low levels of technology, poor infrastructure, low investment in the food production systems, and poor temperature management [56]. Postharvest waste is estimated to be approximately 60% depending on the production region, the season, and the crop [57]. In Brazil, for example, the post-harvest losses and waste of vegetables and fruits is approximately 30% [53]. It can lead to soil fertility challenges, especially where agriculture is predominant. Soil fertility problems occur as a result of inadequate amounts of residues which through direct application, as manure, or as compost find their way back to the land [52]. Examples of agricultural residues include straw from wheat, stover, cobs and corn from maize, husks and shells from coconuts, and stalks from cotton.

Stages of an entire postharvest system include harvesting, threshing, drying, storing, processing, and use. Harvesting operations account for 5–8% losses, storage 15–20% and transport 10–12% [58]. The key sources of postharvest waste in economically developing and economically developed nations are uncontrolled handling, substandard planning of the amount to purchase, and defective packaging [53].

In a study, [52] assessed the residues that are generated on a farm at the time of harvest and also considered the by-products that are generated when crops are processed, for example, sugarcane bagasse produced from sugarcane. The investigation found the total post-harvest losses to be about 92 Mton/year, where 32 Mton/year is attributed to sugarcane losses, 16 Mton/year to wheat losses, and 9 Mton/year to rice.

Studies have pointed out that the reduction of postharvest waste can contribute to increasing the availability of food in the food system, thereby reducing food insecurity, improving farmers' income, bettering nutrition, and reducing the wasting of critical resources such as water, land, energy [57]. Additionally, postharvest waste can be used to generate valuable products such as bioenergy and biochar when technologies such as gasification and anaerobic digestion are employed.

4. Energy resource reclamation

Paper, cardboard, food waste, grass clippings, leaves, wood, and leather goods are examples of biogenic (plant or animal-based) materials. Non-biogenic combustible materials include plastics and other petroleum-based synthetic materials, as well as non-combustible materials like glass and metals. Many nations employ waste-to-energy plants to harness the energy contained in solid waste. Waste-to-energy facilities are widely used in various European nations and Japan, owing to a scarcity of open landfill areas in such countries. Solid waste is often burnt in waste-to-energy facilities, which use the heat from the fire to produce steam, which is then used to generate electricity or heat buildings. **Figure 1** shows the worldwide composition of solid waste from homes, towns, and farms as follows: paper (18%), plastics (12%), organic materials (43%), glass (5%), metal (4%), rubber, leather, and textile (9%), and miscellaneous materials (9%). (9%) [58].

Different types of biofuels may be recovered and purified from organic waste fractions for usage at home or in the workplace. The energy content of garbage determines the quantity of energy that can be retrieved (calorific value). **Figure 2** shows the average energy content of solid waste from houses, towns, and farms, with paper having a potential of 16 MJ/kg, plastics 35 MJ/kg, organics 4 MJ/kg, glass and metal 0 MJ/kg, and other 11 MJ/kg [59].

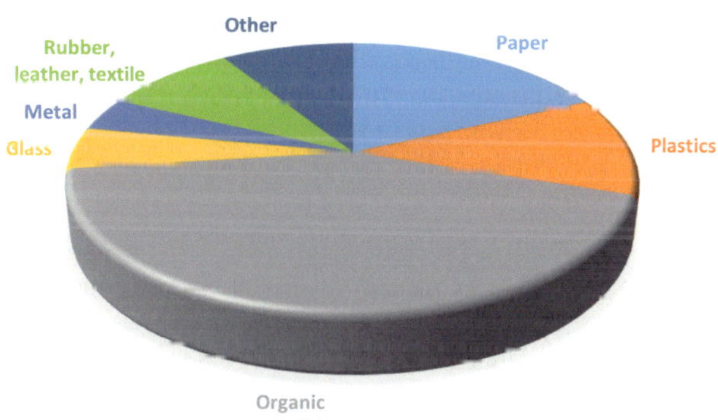

Figure 1.
Typical composition of solid waste from households, municipalities and farms.

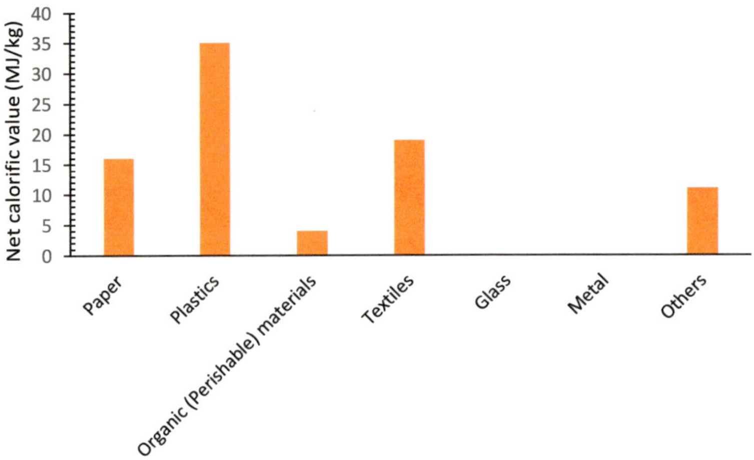

Figure 2.
The average energy content of home/municipal/farm solid waste.

Most biogenic and non-biogenic components may be found in household, municipal, and farm solid waste (HMFSW) used to produce power. Newsprint, paper, cartons/packaging, textiles, wood, food waste, yard trimmings, and leather are biogenic components, while rubber, PET, HDPE, PVC, LDPE/LLDPE, PP, PS, and other (plastic) and metals are non-biogenic [60]. The biogenic fraction of solid waste declines when consumers reuse or recover more biogenic waste (such as paper, packaging, food waste, and yard trimmings) while discarding more non-biogenic trash (such as plastics and metals). Because non-biogenic material has a larger heat content than biogenic material, as shown in **Figure 2**, the average heat content of HMFSW as a whole is rising, making it a more efficient fuel for generating power [59].

Because of a rise in the consumption (and discarding) of non-biogenic materials, as well as enhanced recovery of biogenic materials before they reach the waste stream as discards, the biogenic proportion of HMFSW continues to decline due to more recycling activities. As a result, as the consumption of plastics rises, renewable energy provided by solid garbage decreases, and biogenic waste is more collected and/or repurposed [61].

The technologies that emphasize the biochemical processes leading to the production of biogas in **Figure 3** are discussed in depth in this chapter. The anaerobic generation of biogas, which is a combination of methane and carbon dioxide, was described in detail by Angelidaki et al. [62]. They found three major physiological groups of microorganisms that drive the bio-methanation process: (1) primary fermenting bacteria, (2) anaerobic oxidizing bacteria, and (3) methanogenic archaea, a phylogenetically varied group of strictly anaerobic *Euryarchaeota* whose energy metabolism is limited to the production of methane from carbon dioxide and hydrogen, formate, methanol, methylamines, and/or acetate.

4.1 Bio-hydrogen production

Bio-hydrogen can be produced through different thermochemical, electrochemical and biological processes. By 2020, steam reforming of natural gas, partial oxidation of methane, and coal gasification had produced around 95% of hydrogen from fossil fuels [63]. Other techniques of hydrogen synthesis include biomass gasification, methane pyrolysis with no CO_2 emissions, and water electrolysis. The later processes, such as methane pyrolysis and water electrolysis, may be carried out using any form of electricity, including solar power [64].

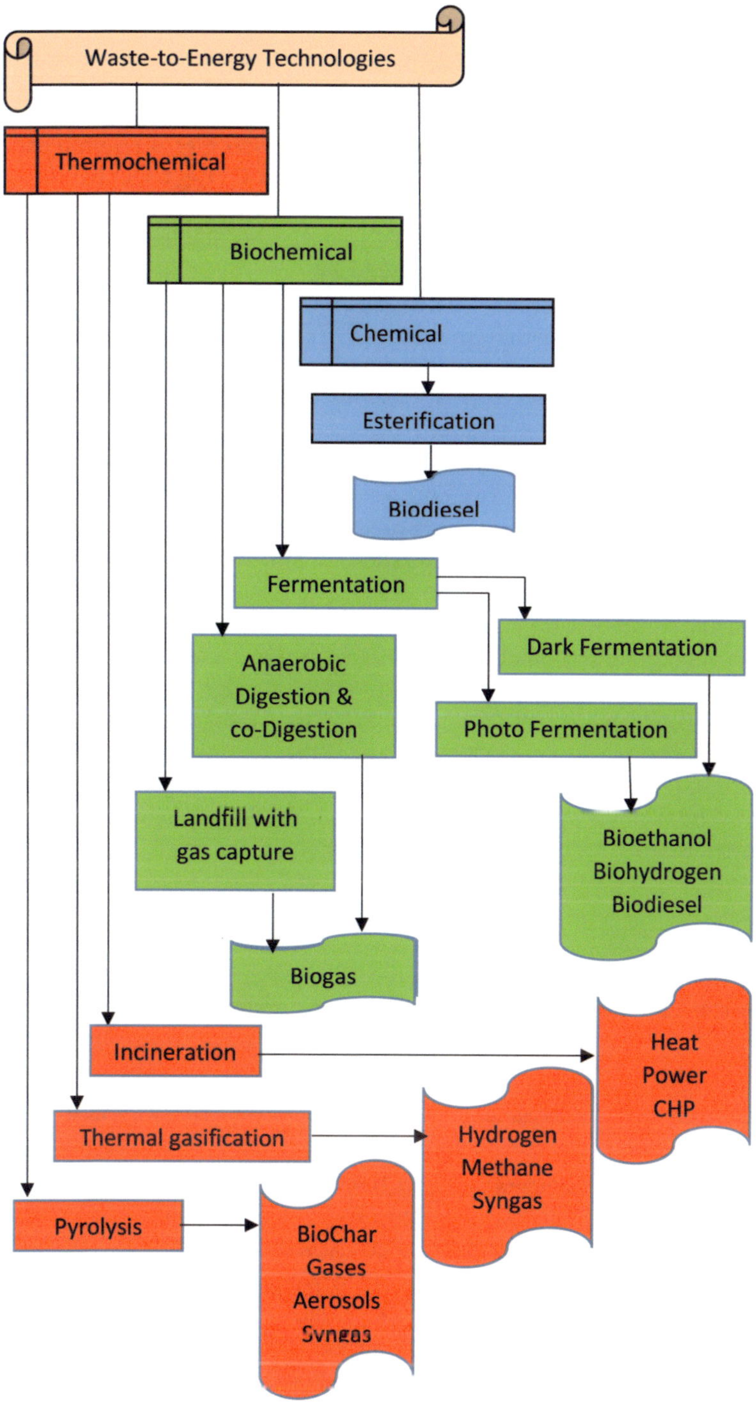

Figure 3.
The technologies that drive the conversion of waste to energy.

Anaerobic and photosynthetic bacteria can produce bio-hydrogen from carbohydrate-rich and non-toxic basic materials. Hydrogen is obtained as a by-product during the conversion of organic wastes into organic acids, which are subsequently utilized to generate methane under anaerobic conditions [63, 64]. The availability, affordability, carbohydrate content, and biodegradability of waste materials utilized in bio-hydrogen generation are the most important factors to consider. Simple sugars like glucose, sucrose, and lactose are easily biodegradable and are excellent hydrogen generation substrates.

Ginkel investigated hydrogen synthesis from industrial effluents from confectioners, apple and potato processors, as well as greywater. From potato processing wastewater, the maximum production yield was 0.21 L H_2/g COD [65, 66].

Bio-hydrogen may be produced in a variety of techniques, including dark fermentation (DF), microbial fuel cells (MFC), and microbial electrolysis cells (MEC). DF and MFC are more effective for bio-hydrogen synthesis from carbohydrate-rich effluents than photo-process (CRE). To enhance the optimum bio-hydrogen generation, either hydrothermal preparation of the inoculum or a high dilution rate can be utilized to minimize the activity of inhibiting bacteria and lower the pH of the medium. For an optimum bio-hydrogen generation, a mixed microbial culture is more trustworthy than a pure microbial source because pure cultures take more care to maintain, but mixed cultures include a wider range of bacteria for the biological conversion of organic materials into useful products. Some of the technologies that produce bio-hydrogen are given in **Table 1**.

Among the various renewable energy sources, bio-hydrogen is gaining a lot of traction as it has very high efficiency of conversion to usable power with less pollutant generation. During fermentation, bacteria release enzymes that hydrolyse biopolymers, resulting in depolymerization of lipids, proteins, nucleic acids, and carbohydrates to intermediate soluble monomers such as fatty acids, glycerol, amino acids, purines, pyrimidines, Mono sugars, and others, which are then converted to short-chain fatty acids, alcohols, hydrogen, and carbon dioxide. **Table 1** shows that different technologies use different fermentation methodologies to transform organic substrates into hydrogen in the absence or presence of light, such as dark fermentation and photo-fermentation. Bio-hydrogen is a valuable and potential source of energy [71].

4.2 Biogas

Agricultural leftovers, such as manure and straw, are among the many potential substrates for AD [72]. However, because of their high percentage of lignocellulose, which is difficult to decompose due to its complicated structure, with cellulose

Waste feedstock	Waste source	Resource recovered	Technology used	Reference
Cheese process effluent	Agro-based industry	Hydrogen	Thermophylic DF	[63]
Sugar mill effluent	Agro-based industry	Bio-electricity	MFC	[63, 65]
Bagasse hydrolysate	Agro-based industry	Bio-hydrogen	DF	[63, 67]
Brewery wastewater	Agro-based industry	Bio-hydrogen	DF	[63, 67, 68]
Olive mill effluent	Agro-based industry	Bio-hydrogen	DF	[63, 69, 70]
Food waste	Household	Bio-hydrogen	Bio-processor	[69]
Greywater	Household	Bio-hydrogen	AD, DF	[70]
Blackwater	Household	Bio-hydrogen	AD, DF	[69, 70]

Table 1.
Technologies that produce bio-hydrogen from waste resources.

fibers securely connected to hemicellulose and lignin, their utility for biogas production is still limited. As a result, lignocellulosic materials have a sluggish decomposition rate and a poor biogas output [72, 73].

4.2.1 Production of biogas from agricultural waste

Around half of the world's habitable land is dedicated to agriculture [74] and is the largest ecosystem managed by humans [75]. Due to continual development and intensification in response to the important dietary needs of the growing populace and bioenergy demand, agriculture has become the most anthropic activity with the largest impact on the environment, especially in the developing countries (ES) [76].

While agricultural landscapes have a lot of potential for reaching renewable energy objectives and supporting local economies, bioenergies are frequently seen as a contentious solution for long-term development due to the rivalry for agricultural land. In recent years, a lot of effort has gone into resolving such food-energy conflicts. A promising source of renewable energy is the use of residual biomass for energy production. It has gained significant economic and environmental importance in recent decades, and it has the potential to close material and energy cycles, protect the environment, recover resources, and reduce the impact and quantity of wastage [77].

Biogas is a versatile biofuel that can be produced from a variety of feedstocks [78]. The anaerobic digestion (AD) process facilitates the transformation of biomass into energy and digestate using biogas technology. The energy generated by (AD) has been utilized to generate heat, electricity, and biomethane, the digestate has also been used as a biofertilizer to restore soil nutrient levels and so boost feedstock productivity [79]. Biogas will account for 25% of all bioenergy in Europe (shortly) due to its many benefits for energy supply, security, and economic benefits [80].

Many agricultural residues have the potential to be valuable resources if they are managed properly. Stalks, straw, leaves, roots, husks, seed shells, and farm and animal farming waste make up the raw material base. These sources of biomass have a diverse set of properties. The most noteworthy difference is between dry residues (such as straw) and those that are more suited to thermo-chemical conversion routes such as combustion, gasification, and pyrolysis, while wet residues (such as animal slurries) which are more suited to biological conversion routes, like biogas production as depicted in **Figure 4** [80].

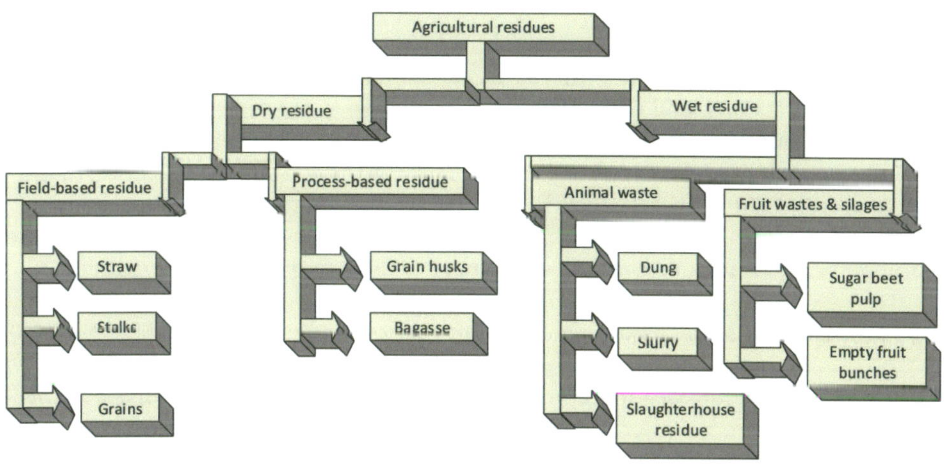

Figure 4.
Classification of agricultural residues.

Various studies into the South African wine industry have looked at grape pomace as a potential biogas feedstock, notably in the Western Cape, which has a concentrated wine sector [77]. The study found that because the wine and grape industry is reliant on seasonal production, the use of grape pomace for energy generation is not feasible for a sole. However, the study found that 1 tonne of grape pomace could produce approximately 230 m^3 of biogas and 828 kWh of renewable electricity. According to the study, communal digesters serving neighboring wineries would increase their viability as a long-term remedy to winery waste [81].

4.2.2 Production of biogas from food waste

A report published by the Food and Agriculture Organization of the United Nations (FAO) in 2019, Globally, over 33% of human food is wasted, equivalent to approximately 1.3 billion tonnes each year. Food waste per capita in West Asia and North Africa amounts to 6–11 kg per annum, compared to 95–115 kg in Western countries. Food waste occurs throughout the food supply chain, including agricultural processing, sorting, storage, transportation, distribution, selling, preparation, cooking, and serving [82]. Food waste costs the world economy over $ 750 billion (US) annually [83].

Compared to other technologies such as incineration and landfilling, AD of food wastes has a lower environmental impact [82, 84]. As a result, multiple efforts to enhance biogas production from food waste have been made in recent years [85]. Despite the high potential for valorising waste food into biogas in nearly any city on the planet, not many industrial-scale plants, especially in industrialized countries, have been put into service [82].

4.3 Biochar

Crop residues, non-commercial wood and wood waste, manure, solid waste, non-food energy crops, construction scraps, yard trimmings, methane digester residues, or grasses are used in the production of sustainable biochar. Biomass for biofuels or biochar must be surplus that is, more than what should be left on-site to maintain forest and agricultural cropland health [86].

Biochar is created when biomass is pyrolyzed or gasified. These are thermal conversion methods that involve superheating and thermally converting biomass at high temperatures (350–700°C) in a specially designed furnace that captures all of the emissions produced [87].

Biochar is just one of the many valuable bioenergy and bioproducts produced during pyrolysis. Volatile gases (methane, carbon monoxide, and other combustible gases), hydrocarbons, and the majority of the oxygen in the biomass are burned or driven off, resulting in carbon-enriched biochar. All of the emissions (also known as air pollution and greenhouse gases) produced by burning biomass are captured and condensed into liquid fuels such as bio-oil, industrial chemicals, or syngas (synthetic gas). These products can be containerized for sale, stored for future use at the manufacturing facility, or used on-site as part of the energy production process.

4.3.1 Production of biochar from agricultural waste

Biochar is a unique carbonaceous porous material generated by pyrolysis or thermochemical conversion of biomass with little or no oxygen [88]. Due to its distinct characteristics such as large surface area, porous structure, oxygenated functional groups, and cation exchange capacity, biochar has recently attracted increased attention in several engineering applications [89].

Biochar can be made from a variety of different feedstocks, wood chips and pellets, tree bark, crop residues (corn stover, nutshells and rice hulls), and other feedstocks that are used commercially, internationally and in research studies. Organic wastes such as grain, sugarcane bagasse, chicken and dairy manure, and sewage sludge have been studied as potential feedstocks [90–96].

As a result, BC has a huge diversity of composition. Xie et al. [97] compiled a list of biochar conversion technologies, detailing product yields and operating conditions, finding that the biochar yield ranged from 15 to 35% with long residence periods of up to 4 h at a moderate temperature of not more than 500°C, and the bio-oil yield ranged from 30 to 50%. More bio-oil (50–70%) was discovered with a shorter residence time (up to 2 s). The thermochemical processes of pyrolysis and carbonization are used to convert biomass into biofuels and other bioenergy products. Biochar is produced by pyrolysis, thermochemically converts biomass in the absence of oxygen at a temperature greater than 400°C. The main components of biochar are carbon (C), hydrogen (H), oxygen (O), nitrogen (N), sulfur (S), and ash. Pyrolysis is divided into three categories: slow, intermediate, and rapid. Kung et al. [98] found a slow pyrolysis process produced more biochar than other processes. Steiner et al. [99] used a top-lit updraft gasifier to make biochar from rice husk and discovered farmers produce biochar in the field with a 15–33% efficiency. Each year, biochar made from on-farm crop residues can contribute 6.3–11.8% of the total production area [100]. Carbonization (a slow pyrolysis process) produces biochar as a by-product and has been around for thousands of years. Slow pyrolysis is a technique for heating biomass to a low temperature (400°C) in the absence of oxygen over a long period [99].

4.3.2 Energy recovery from the production of biochar

Two aspects of biochar production's energy recovery have been identified. The first is that the energy value of the steam, gas, and oil by-products of biochar production can be recovered, resulting in a secondary revenue stream and a reduction in greenhouse gas (GHG) emissions [100, 101]. The volatiles in the feedstock burned during pyrolysis, releasing energy as heat, which can be used to generate steam or for combustion in electricity generation plants [102]. Bio-oils can be refined into transportation fuels or burned to provide energy for heating if adequate quantities are available [103]. The syngas and bio-oils can be used to generate steam, which can then be used to power turbines in centralized power plants [103]. However, the ability to use bio-oil is limited by the size of the operation and the volume of oils produced.

Secondly, biochar can be burned directly as a carbon-neutral or low-carbon energy source when. Worldwide, 41 million tonnes of char are produced annually for cooking and industrial purposes [104].

Biochar production and use consume less energy than burning wood for cooking or heating directly [105]. Because it expands the feedstock base to include crop residues and other by-products of agricultural-related activities. These feedstocks are already being used to meet a large portion of the world's household energy needs [106]. When opposed to conventional cooking fuels (e.g. paraffin) which pose indoor fire risks and health problems linked with poor indoor air quality due to its combustion, using char for energy provides several social benefits [107].

4.4 Charcoal

In 2018, coal combustion accounted for about 38% of global electricity production [108]. According to the International Energy Agency (IEA), the world's

Waste feedstock	Source of waste	Energy resource	Technology	Biogas potential (m3/ kg dry mass)	Reference
Food waste	Household	Biogas	AD	0.027–0.312	[72, 84, 87]
Animal dung	Agro-waste	Biogas	AD	0.012–8.25	[76, 78]
Bagasse	Agro-based waste	Biogas	AD	0.182	[72, 76, 79]
Blackwater	Household	Biogas	AD	0.052–0.232	[82, 83, 87]
Greywater	Household	Biogas	AD	0.035–0.145	[82, 83, 87]
Silage	Agro-waste	Biogas	AD	0.213–0.458	[72, 77, 79]
Husks	Agro-based waste	Biogas	AD	0.013–0.	[77, 79]
Slaughterhouse residue	Agro-based waste	Biogas	AD	0.315–0.812	[76, 87]
Straw	Agro-waste	Biogas	AD	0.161–0.214	[72, 83, 85]
Municipal waste	Household	Biogas	AD	0.035–0.268	[76, 84, 87]

Table 2.
The potential of energy resource extraction from household and agricultural wastes.

recoverable coal reserves are roughly 888.9 billion tonnes, with the majority of them situated in China, Australia, India, Russia, South Africa, and The United States of America [108]. The IEA's energy supply statistics for coal in 2003 and 2017 were 2,619,947 kilo tonnes and 3,789,934 kilo tonnes, respectively. Annual global combustion was estimated to be around 2.5 billion tons [109].

Charcoal is a carbon-rich solid that is derived from biomass in the same way. Charcoal is typically used for heating or cooking and is associated with barbecuing. The temperature at which charcoal and biochar are produced is a significant difference. Charcoal is produced at temperatures ranging from 400 to 1000°C, whereas biochar is produced at temperatures ranging from 600 to 1000°C. When biochar is made at lower temperatures, volatiles (smokiness) are left behind, which has been shown to limit plant growth [105].

The temperature affects porosity as well; the higher the temperature, the greater the porosity. This means that charcoal is not as good as biochar at retaining water and nutrients. Microbes have less surface area when there are fewer pores. As a result, using crushed up charcoal instead of biochar will not be as beneficial to your plants. Because charcoal is made at a lower temperature, it produces a less stable form of carbon, which means it does not provide the long-term carbon sequestration properties associated with biochar [99].

Carbon, which occurs naturally in wood, forms a crystalline structure at higher temperatures, according to research. That is, charcoal has a shorter lifespan in the soil than biochar, which can last hundreds, if not thousands, of years. As a result, it is less effective in the soil and less beneficial to the environment. Biochar made at higher temperatures performs better and sequesters more carbon [87]. The potential of extracting biogas from waste resources is shown in **Table 2.**

5. Economic impact of resource reclamation

Population growth resulted in modest increases in per capita income at the macro level. Economic activity in any of the world's poorest countries, on the other hand, has stalled. This economic downturn coincided with significant (and, in some cases,) population growth, resulting in stagnant or decreasing per capita incomes [1]. Agricultural economics applies economic ideas to agriculture without taking into account the profession's economic, social, and environmental concerns. It's important to remember that agricultural economics encompasses a far larger spectrum of food and fiber-related activities than just farming. The agriculture sector accounts for around 12–15% of the nation's production when considered in this light [110].

The circular economy notion [111] is a valuable method of comprehending the waste management hierarchy's implementation in terms of its contribution to the green economy and other energy power recovery (EPR) initiatives. A circular economy is characterized as "closing the loop" between resource extraction and waste disposal throughout the economic cycle by applying waste avoidance, reuse, repair, recycling, and recovery strategies to minimize waste output and demand for virgin resources as production inputs. An economy that is meant to be restorative and regenerative to maintain the greatest utility and value of goods, components, and materials [112].

Recycling efforts account for a significant portion of waste reclamation. In a circular economy, the interchange of products and services is boosted. As the activities in the sector increase, the economic sectors of interest are impacted. This affects the income of all active participants in the economy. Cans and bottles may be recycled by consumers, shipping cardboard and unsold food can be recycled by businesses, and scrap materials can be recycled by manufacturers. Thousands of recycling brokers and processors exchange source-separated and aggregated materials, as well as treat waste to provide feedstock for manufacturers to employ as product inputs. This shows that waste recovery initiatives have a substantial global economic impact.

6. Life cycle analysis (LCA) of waste resource reclamation

The propensity of life is based on the creation of new cells and the elimination of old or expired ones. Another way of looking at life is as a process of ingesting nutrients-based materials and expelling waste. As a result, waste generation is a normal occurrence. Man's effort to bring rationality to waste generation and disposal procedures is known as waste management. LCA is defined by the Society of Environmental Toxicology and Chemistry (SETAC) [113] as "an objective process for evaluating the environmental burdens associated with a product, process, or activity, by identifying and quantifying energy and materials used, waste released to the environment, and evaluating and implementing opportunities to effect environmental improvements." LCA is a methodology for analyzing the environmental implications of a product, process, or service from the raw material production through the final disposal of wastes.

6.1 Composition of household and farm wastes

The composition of HMFSW from households, municipalities, and farms is influenced by a variety of factors including cultural traditions, lifestyles, eating preferences, climate, and income. Many diverse sources of solid waste were found by Yadav and Samadder [114] in families, municipalities, and crop farms. Family units, hostels, governmental and private organizations, and commercial centres all produce waste. Waste is created on the farm during harvest and post-harvest operations. Solid wastes can be categorized into biogenic solid waste (BSW) and non-biogenic solid waste (nBSW) based on their origins. Location, socioeconomic position, habits, environmental awareness, and other variables all influence the frequency of one over the other [113]. The biogenic constituent of solid waste includes paper, packaging/cartons, wood, textiles, food leftovers and waste, yard trimmings, leather, and others; while the nBSW include rubber, polyethene tetrafluoride (PET), high-density polyethene (HDPE), polyvinyl chloride (PVC), low-density polyethene (LDPE), polypropylene (PP), polystyrene (PS), and other plastics. The biogenic component supports biogas production after some customized pretreatment steps.

The Life Cycle Analysis (LCA) process is most commonly utilized as a support tool in the strategic planning and decision-making process for Waste-to-Energy projects [113]. However, the Waste-to-Energy systems' inputs and outputs differ from one project to another; in particular, the waste composition and cost are highly dependent on the project's location. The WtE plant design and waste composition can have a considerable impact on efficiency and emissions.

6.2 Flow characterization of materials

Any country's waste management industry is under growing pressure to improve its environmental performance. Solid waste management (SWM) is essentially a local responsibility in most countries [115]. Low- and middle-income nations confront hurdles in terms of sustainable waste management strategies compared to higher-income countries due to a lack of resources and ability in local governments, as well as the ineffective execution of specialized regulations. As a result, nations with greater incomes are leading the way in creating sustainable waste management systems. Source reduction is the most prevalent waste management method in the sustainable waste management ladder.

In a stated mapping strategy, the material flow analysis (MFA) technique is utilized to characterize or quantify the efficiency of waste collection and disposal. Due to their complexity and volume, the system is divided into four subsystems to reflect the management of major waste streams: residual trash, commingled materials, source segregated dry recyclables, and source segregated food and garden wastes. The collected primary waste streams are the system's import flows, while secondary products and emissions are the system's export flows [116].

The use of integrated material flow analysis (MFA) and life cycle analysis (LCA) to make decisions in SWM systems is becoming increasingly popular. By acting as a great tool for assessing and controlling flows of wastes, secondary products, and residues, MFA on the levels of commodities aids in understanding the functioning of processes and the connections between processes in waste management [117]. LCA assesses the environmental advantages and disadvantages of waste management solutions. LCA examines system performance and allows for alternative comparisons as well as the identification of potential system improvements.

7. Conclusion

The increase in human population and the rural-urban drift has continuously placed a strain on fossil energy resources. Waste piles have increased across the globe with limited land space for landfills. Solid waste from the household, municipality and farm have an inherent energy content that could be harnessed to bridge the energy gap that has continued to get wider due to the increasing demand. The biogenic component of the solid waste is sorted for customized biochemical processes thereby accessing energy resources that could be used in homes or private and public institutions. Anaerobic digestion produces methane and carbon dioxide. However, the system could be tailored to produce hydrogen, which is an energy resource of value. The circular economy is necessarily employed to cub the waste piles and to enhance environmental sustainability.

One of the methods to ensure equilibrium in the energy economy is converting waste resources into value materials. Bioethanol, biogas, biohydrogen are some of the energy resources that can be extracted from the BSW; and because BSW is constantly produced from their sources, these energy resources are described as renewable. The residue that is left after extracting the energy resources can be turned into biochar, which is a resource that could be used to amend the soil for agricultural production.

Acknowledgements

The authors appreciate the support of the Durban University of Technology for funding this research focus area in Water and Wastewater. This work is also supported by the Green Engineering and Sustainability Group. The authors would like to acknowledge enabling atmosphere provided to harness the scientific information compiled and presented in this book chapter.

Conflict of interest

The authors declare that there is no conflict of interest.

Nomenclature

AD	anaerobic digestion
BSW	biogenic solid waste
EPR	energy and power recovery
FAO	Food and Agriculture Organization
GHG	greenhouse gas
HDPE	high-density polyethene
HMFSW	household, municipal and farm slid waste
IWWT	Institute of Water and Wastewater Technology
LCA	life cycle analysis
LDPE	low-density polyethene
LLDPE	linear low density polyethelene
MFA	material flow analysis
nBSW	nonbiogenic solid waste
PET	polyethylene terephthalate
PP	polypropylene
PS	polystyrene
SETAC	Society of Environmental Toxicology and Chemistry
SWM	solid waste management
USAB	up-flow anaerobic sludge blanket

Author details

Donald Kukwa[1*], Maggie Chetty[1], Zikhona Tshemese[2], Denzil Estrice[1] and Ndumiso Duma[3]

1 Department of Chemical Engineering, Durban University of Technology, Durban, South Africa

2 Department of Chemistry, Durban University of Technology, Durban, South Africa

3 Department of Chemical Engineering, University of KwaZulu-Natal, Durban, South Africa

*Address all correspondence to: tyokedroo@gmail.com

References

[1] Adamcová D, Vaverková MD, Stejskal B, Břoušková E. Household solid waste composition focusing on hazardous waste (in English). Polish Journal of Environmental Studies. 2016;**25**(2):487-493. DOI: 10.15244/pjoes/61011

[2] Rousta K. Household Waste Sorting at The Source: A Procedure for Improvement. Sweden: PhD, University of Borås, Boras; 2018. [Online]. Available from: http://urn.kb.se/resolve?urn=urn:nbn:se:hb:diva-12821

[3] Salihu A. Public perception on solid waste management and related government policy. Texila International Journal of Public Health. 2018;**6**(1):1-9. DOI: 10.21522/TIJPH.2013.06.01.Art001

[4] Zhou X. Resources overview of solid waste. In: Presented at the 6th International Conference on Machinery, Materials, Environment, Biotechnology and Computer (MMEBC). 2016

[5] Sinha RK, Valani D, Sinha S, Singh S, Herat S. Bioremediation of contaminated sites: A low-cost nature's biotechnology for environmental cleaning up by versatile microbes, plants & earthworms. In: Faerber T, Herzog J, editors. Solid Waste Management and Environmental Remediation. Nova Science Publishers, Inc.; 2009. p. 72. (Ch. 1)

[6] Reddy PJ. Municipal Solid Waste Management: Processing, Energy Recovering and Global Examples. Hyderabad, India: BS Publications; 2011. pp. 1-449

[7] Viljoen JMM, Schenck CJ, Volschenk L, Blaauw PF, Grobler L. Household waste management practices and challenges in a rural remote town in the Hantam Municipality in the Northern Cape, South Africa, (in English). Sustainability. 2021;**13**(11):5903. DOI: ARTN 5903_10.3390/su13115903

[8] Ajayi AP, Sambo IS. Assessment of household solid waste management in Gombe metropolis. IOSR Journal of Environmental Science, Toxicology and Food Technology. 2021;**15**(8 Ser. I):1-6. DOI: 10.9790/2402-1508010106

[9] Abdul-Rahman F. Reduce, Reuse, Recycle: Alternatives for Waste Management. New Mexico State University. Available from: aces.nmsu.edu/pubs/_g/G314.pdf [Accessed: 2021]

[10] Carnevale C, Gabusi V, Volta M. POEM-PM: An emission model for secondary pollution control scenarios, (in English). Environmental Modelling & Software. 2006;**21**(3):320-329. DOI: 10.1016/j.envsoft.2004.11.003

[11] Suthar S, Singh P. Household solid waste generation and composition in different family size and socio-economic groups: A case study. Sustainable Cities and Society. 2015;**14**:56-63

[12] Dangi MB, Pretz CR, Urynowicz MA, Gerow KG, Reddy J. Municipal solid waste generation in Kathmandu, Nepal. Journal of Environmental Management. 2011;**92**(1):240-249

[13] Okalebo S, Opata G, Mwasi B. An analysis of the household solid waste generation patterns and prevailing management practices in Eldoret town, Kenya. International Journal of Agricultural Policy and Research. 2014;**2**(2):076-089

[14] Philippe F, Culot M. Household solid waste generation and characteristics in Cape Haitian city, Republic of Haiti. Resources, Conservation and Recycling. 2009;**54**(2):73-78

[15] Oguntoke O, Amaefuna B, Nwosisi M, Oyedepo S, Oyatogun M. Quantification of biodegradable household solid waste for biogas production and the challenges of waste sorting in Abeokuta Metropolis, Nigeria.

International Journal of Energy and Water Resources. 2019;**3**(3):253-261

[16] Sujauddin M, Huda S, Hoque AR. Household solid waste characteristics and management in Chittagong, Bangladesh. Waste Management. 2008;**28**(9): 1688-1695

[17] Ojeda-Benítez S, Armijo-de Vega C, Marquez-Montenegro MY. Household solid waste characterization by family socioeconomic profile as unit of analysis. Resources, Conservation and Recycling. 2008;**52**(7):992-999

[18] Koushki P, Al-Khaleefi A. An analysis of household solid waste in Kuwait: Magnitude, type, and forecasting models. Journal of the Air & Waste Management Association. 1998;**48**(3):256-263

[19] Qu X-Y, Li Z-S, Xie X-Y, Sui Y-M, Yang L, Chen Y. Survey of composition and generation rate of household wastes in Beijing, China. Waste Management. 2009;**29**(10):2618-2624

[20] Dennison G, Dodd V, Whelan B. A socio-economic based survey of household waste characteristics in the city of Dublin, Ireland—II. Waste quantities. Resources, Conservation and Recycling. 1996;**17**(3):245-257

[21] Ojeda-Benitez S, de Vega CA, Ramırez-Barreto ME. Characterization and quantification of household solid wastes in a Mexican city. Resources, Conservation and Recycling. 2003;**39**(3):211-222

[22] Sha'Ato R, Aboho S, Oketunde F, Eneji I, Unazi G, Agwa S. Survey of solid waste generation and composition in a rapidly growing urban area in Central Nigeria. Waste Management. 2007;**27**(3):352-358

[23] Hahladakis JN, Iacovidou E. An overview of the challenges and trade-offs in closing the loop of post-consumer plastic waste (PCPW): Focus on recycling. Journal of Hazardous Materials. 2019;**380**:120887

[24] Glass B. Maximising the Recyclability of Glass Packaging. Sheffield, UK: British Glass; 2019

[25] Raharjo S, Komala PS, Pratoto A. Biogas production from household solid waste-an alternative solid waste treatment for a communal scale. In: MATEC Web of Conferences. Vol. 276. EDP Sciences; 2019. p. 06016

[26] Szala B, Paluszak Z. Validation of biodegradable waste composting process based on the inactivation of Salmonella senftenberg W775. Polish Journal of Environmental Studies. 2008;**17**(1)

[27] Olanrewaju O, Ilemobade A. Waste to wealth: A case study of the Ondo State integrated wastes recycling and treatment project, Nigeria. European Journal of Social Sciences. 2009;**8**(1):7-16

[28] Achinas S, Achinas V, Euverink GJW. A technological overview of biogas production from biowaste. Engineering. 2017;**3**(3):299-307

[29] Ghosh P et al. Assessment of methane emissions and energy recovery potential from the municipal solid waste landfills of Delhi, India. Bioresource Technology. 2019;**272**:611-615

[30] Khan EU, Mainali B, Martin A, Silveira S. Techno-economic analysis of small scale biogas based polygeneration systems: Bangladesh case study. Sustainable Energy Technologies and Assessments. 2014;**7**:68-78

[31] Liu D, Liu D, Zeng RJ, Angelidaki I. Hydrogen and methane production from household solid waste in the two-stage fermentation process. Water Research. 2006;**40**(11):2230-2236

[32] Kadjo BS, Sangaré D, Sako KM, Coulibaly L. Biogas production from household solid waste by anaerobic

batch reactor. In: 2020 5th International Conference on Renewable Energies for Developing Countries (REDEC). IEEE; 2020. pp. 1-5

[33] Dowling M. Defining and classifying hazardous wastes. Environment: Science and Policy for Sustainable Development. 1985;**27**(3):18-41

[34] Robertson DK et al. Liquid household hazardous wastes in the United States: Identification, disposal, and management plan. Environmental Management. 1987;**11**(6):735-742

[35] Gurbuz IB, Ozkan G. Consumers' knowledge, attitude and behavioural patterns towards the liquid wastes (cooking oil) in Istanbul, Turkey. Environmental Science and Pollution Research. 2019;**26**(16):16529-16536

[36] Syed S. Solid and liquid waste management. Emirates Journal for Engineering Research. 2006;**11**(2):19-36

[37] Omani JA. Assessment of household liquid waste management a case study of Accra, Ghana. Quantum Journal of Engineering, Science and Technology. 2021;**2**(2):16-26

[38] Cox J, Giorgi S, Sharp V, Strange K, Wilson DC, Blakey N. Household waste prevention—A review of evidence. Waste Management & Research. 2010;**28**(3) z:193-219

[39] Papin V, Bezuglov R, Veselovskaya E, Yanuchok A. Logistics and principles of use of solid and liquid household waste, as relevant energy resource. In: E3S Web of Conferences. Vol. 126. EDP Sciences; 2019. p. 00011

[40] Roshchin A, Grigoriev V, Sagittarius A, Nikolaev A, Rayevskaya E, Usin V. Supercritical hydrothermal destruction of the dangerous substances containing organic chemistry and waste with the prospect of utilization of energy potential of gaseous products. Chemical Physics. 2017;**36**(7):18-24

[41] Puspawati S, Soesilo T, Soemantojo R. An overview of biogas utilization from tempeh wastewater. In: IOP Conference Series: Earth and Environmental Science. Vol. 306(1). IOP Publishing; 2019. p. 012019

[42] Kumar SA, Marimuthu C, Balaji E, Riswan SS. Biogas production from kitchen waste water using USAB reactor. International Journal of ChemTech Research. 2014;**6**(9):4135-4142

[43] Tasnim F, Iqbal SA, Chowdhury AR. Biogas production from anaerobic co-digestion of cow manure with kitchen waste and Water Hyacinth. Renewable Energy. 2017;**109**:434-439

[44] Rycroft M. Municipal liquid waste: A neglected source of energy. Sustainable Energy. 2013:47-49

[45] CSGN. Introduction to annual and perennial plants. In: Creating and Sustaining Your School Garden Spring. 2012. Available from: www.csgn.org/csysg%0ASection

[46] Vico G, Brunsell NA. Tradeoffs between water requirements and yield stability in annual vs. perennial crops. Advances in Water Resources. 2018;**112**:189-202. DOI: 10.1016/J.ADVWATRES.2017.12.014

[47] Borém A, Doe JA, Diola V. Molecular biology and biotechnology. Sugarcane: Agricultural Production, Bioenergy and Ethanol. 2015:257-274. DOI: 10.1016/B978-0-12-802239-9.00012-8

[48] Zimdahl R. Weed-management systems. In: Fundamentals of Weed Science. 5th ed. Academic Press; 2018. pp. 609-649

[49] Hatfield JL, Walthall CL. Climate change: Cropping system changes and adaptations. In: Encyclopedia of Agriculture and Food Systems. 2014

[50] Obi FO, Ugwuishiwu BO, Nwakaire JN. Agricultural waste

concept, generation. Utilization and Management. 2016;**35**(4):957-964

[51] Ungerer M, Bowmaker-Falcone A, Oosthuizen C, Phehane V, Strever A. The future of the Western Cape agricultural sector in the context of the Fourth Industrial Revolution. 2018. [Online]. Available from: https://www.usb.ac.za/wp-content/uploads/2018/07/THE-FUTURE-OF-THE-WC-AGRICULTURAL-SECTOR-IN-THE-CONTEXT-OF-4IR-FINAL-REP.pdf [Accessed: September 14, 2021]

[52] Cardoen D, Joshi P, Diels L, Sarma PM, Pant D. Agriculture biomass in India: Part 2. Post-harvest losses, cost and environmental impacts. Resources, Conservation and Recycling. 2015;**101**:143-153. DOI: 10.1016/J.RESCONREC.2015.06.002

[53] dos Santos SF et al. Post-harvest losses of fruits and vegetables in supply centers in Salvador, Brazil: Analysis of determinants, volumes and reduction strategies. Waste Management. 2020;**101**:161-170. DOI: 10.1016/J.WASMAN.2019.10.007

[54] Chibuzor M. Feed the World—Top 10 Ways of Eliminating Post-Harvest Wastes. Agro4Africa; 2021. Available from: https://agro4africa.com/eliminating-post-harvest-wastes/ [Accessed: September 14, 2021]

[55] Shipman EN, Yu J, Zhou J, Albornoz K, Beckles DM. Can gene editing reduce postharvest waste and loss of fruit, vegetables, and ornamentals? Horticulture Research. 2021;**8**(1):1-21. DOI: 10.1038/s41438-020-00428-4

[56] Gustavsson J, Cederberg C, Sonesson U, van Otterdijk R, Meybeck A. Global Food Losses and Food Waste. Rome; 2011 [Online]. Available from: http://www.fao.org/fileadmin/user_upload/suistainability/pdf/Global_Food_Losses_and_Food_Waste.pdf [Accessed: September 15, 2021]

[57] Yahia EM, Fonseca JM, Kitinoja L. Postharvest losses and waste. Postharvest Technology of Perishable Horticultural Commodities. 2019:43-69. DOI: 10.1016/B978-0-12-813276-0.00002-X

[58] Grolleaud M. Post-harvest Losses: Discovering the Full Story. FAO; 2002. Available from: http://www.fao.org/3/ac301e/AC301e00.htm [Accessed: September 14, 2021]

[59] DOE (U.S. Department of Energy). National algal biofuels technology review. In: Energy Efficiency & Renewable Energy. U.S. Department of Energy, Office of Energy Efficiency and Renewable Energy, Bioenergy Technologies Office; 2016

[60] Dewi Y, Moh F, Adi K, Solimun. Significant factors impact of reclamation on environmental, economic, and social cultural aspects in Makassar city. Global Journal of Engineering and Technology Advances. 2021;**7**(3):213-223. DOI: 10.30574/gjeta.2021.7.3.0094

[61] Bocher BTW, Cherukuri K, Maki JS, Johnson M, Zitomer DH. Relating methanogen community structure and anaerobic digester function. Water Research. 2015;**70**:425-435. DOI: 10.1016/j.watres.2014.12.018

[62] Angelidaki I, Karakashev D, Batstone DJ, Plugge CM, Stams AJM. Biomethanation and its Potential. In: Methods in Enzymology. Vol. 494. Elsevier Inc.; 2011. pp. 327-351. (Ch. 16)

[63] Kumari D, Bhatia RK, Singh R. Technologies and challenges to recover energy products from carbohydrate rich effluents: A mini review. International Journal of Plant, Animal and Environmental Sciences. 2021;**11**(3):423-442. DOI: 10.26502/ijpaes.202113

[64] Rao P, Rathod V. Valorization of food and agricultural waste: A step towards greener future. Chemical Record. 2018;**18**(9):1-15. DOI: 10.1002/tcr.201800094

[65] Goldman G, Ogishi A. Economic impact of waste disposal and diversion in California. In: A Report to the California Integrated Waste Management Board. University of California, Berkeley: Department of Agricultural and Resource Economics; 2001

[66] Sun L, Pope PB, Eijsink VGH, Schnürer A. Characterization of microbial community structure during continuous anaerobic digestion of straw and cow manure. Microbial Biotechnology. 2015;**8**(5):815-827. DOI: 10.1111/1751-7915.12298

[67] Singh A, Sevda S, Reesh IMA, Vanbroekhoven K, Rathore D, Pant D. Biohydrogen production from lignocellulosic biomass: Technology and sustainability, (in English). Energy. 2015;**8**(11):13062-13080. DOI: 10.3390/en81112357

[68] Mohan SV, Chandrasekhar K, Chiranjeevi P, Babu PS. Biohydrogen production from wastewater. In: Pandey A, Chang JS, Hallenbeck PC, Larroche C, editors. Biohydrogen. Burlington: Elsevier; 2013. pp. 223-257. (Ch. 10)

[69] Mohan SV, Srikanth S, Dinakar P, Sarma PN. Photo-biological hydrogen production by the adopted mixed culture: Data enveloping analysis. International Journal of Hydrogen Energy. 2008;**33**:559-569. DOI: 10.1016/j.ijhydene.2007.10.033

[70] Jeguirim M et al. Olive mill wastewater: From a pollutant to green fuels, agricultural water source and bio fertilizer. Biofuel production, (in English). Renewable Energy. 2020;**149**:716-724. DOI: 10.1016/j.renene.2019.12.079

[71] Zhu X et al. Metabolic dependencies govern microbial syntrophies during methanogenesis in an anaerobic digestion ecosystem. Microbiome. 2020;**8**(1):8-22. DOI: 10.1186/s40168-019-0780-9

[72] Tsavkelova EA, Netrusov AI. Biogas production from cellulose-containing substrates: A review. Prikladnaia Biokhimiia i Mikrobiologiia. 2012;**48**(5):469-483. (Russian)

[73] Tubiello F, Salvatore M, Rossi S, Ferrara A, Fitton N, Smith P. The FAOSTAT database of greenhouse gas emissions from agriculture. Environmental Research Letters. 2013;**8**(1):015009. DOI: 10.1088/1748-9326/8/1/015009

[74] Smith S. The role of social capital in the industrialization of the food system: Comment. Agricultural and ResourcLe Economics Review. 2002;**31**(1):25-27

[75] Collier C, de Almeida Neto M, de Almeida G, Rosa Filho J, Severi W, El-Deir A. Effects of anthropic actions and forest areas on a neotropical aquatic ecosystem. Science of the Total Environment. 2019;**691**:367-377

[76] Barros M, Salvador R, de Francisco A, Piekarski C. Mapping of research lines on circular economy practices in agriculture: From waste to energy. Renewable and Sustainable Energy Reviews. 2020;**131**:1099L58

[77] Panoutsou C, Perakis C, Elbersen B, Zheliezna T, Staritsky I. Assessing potentials for agricultural residues. In: Modeling and Optimization of Biomass Supply Chains. Elsevier Inc.; 2017. pp. 169-197. (Ch. 7)

[78] Lijó L et al. Life cycle assessment of electricity production in Italy from anaerobic co-digestion of pig slurry and energy crops. Renewable Energy. 2014;**68**:625-635

[79] Barz M. Agricultural residues—A promising source for bioenergy production. In: Joint International Workshop on Technical and Economic Aspects of Renewable Energies—Know-How Transfer as Development Opportunity for Southern Europe.

Berlin: Mensch und Buch Verlag; 2016. pp. 82-90. ISBN 978-3-86387-721-7

[80] Dillon C. Waste management in the South African wine industry [dissertation]. Partial Requirement for the Diploma of Cape Wine Master, Cape Wine Academy; 2011

[81] Peixoto M, Pinto HS. Desperdício de Alimentos: Questões Socioambientais. (econômicas e regulatórias). 2016

[82] Aghbashlo M, Tabatabaei M, Soltanian S, Ghanavati H, Dadak A. Comprehensive exergoeconomic analysis of a municipal solid waste digestion plant equipped with a biogas genset. Waste Management. 2019;**87**:485-498

[83] Campuzano R, González-Martínez S. Characteristics of the organic fraction of municipal solid waste and methane production: A review. Waste Management. 2016;**54**:3-12

[84] Hagman L, Blumenthal A, Eklund M, Svensson N. The role of biogas solutions in sustainable biorefineries. Journal of Cleaner Production. 2018;**172**:3982-3989

[85] Mao C, Feng Y, Wang X, Ren G. Review on research achievements of biogas from anaerobic digestion. Renewable and Sustainable Energy Reviews. 2015;**45**:540-555

[86] Mumme J, Srocke F, Heeg K, Werner M. Use of biochars in anaerobic digestion. Bioresource Technology. 2014;**164**:189-197

[87] Sunyoto N, Zhu M, Zhang Z, Zhang D. Effect of biochar addition on hydrogen and methane production in two-phase anaerobic digestion of aqueous carbohydrates food waste. Bioresource Technology. 2016;**219**:29-36

[88] Yaman S. Pyrolysis of biomass to produce fuels and chemical feedstocks. Energy Conversion and Management. 2004;**45**(5):651-671

[89] Das K, Garcia-perez M, Bibens B, Melear N. Slow pyrolysis of poultry litter and pine woody biomass: Impact of chars and bio-oils on microbial growth. Journal of Environmental Science and Health, Part A. 2008;**43**(7):714-724

[90] Favreau G et al. Land clearing, climate variability, and water resources increase in semiarid southwest Niger: A review. Water Resources Research. 2009;**45**(7)

[91] Shinogi Y, Yoshida H, Koizumi T, Yamaoka M, Saito T. Basic characteristics of low-temperature carbon products from waste sludge. Advances in Environmental Research. 2003;**7**(3):661-665

[92] Xie T, Reddy K, Wang C, Yargicoglu E, Spokas K. Characteristics and applications of biochar for environmental remediation: A review. Critical Reviews in Environmental Science and Technology. 2015;**45**(9): 939-969

[93] Kung C, Kong F, Choi Y. Pyrolysis and biochar potential using crop residues and agricultural wastes in China. Ecological Indicators. 2015;**51**:139-145

[94] Steiner C et al. Participatory trials of on-farm biochar production and use in Tamale, Ghana. Agronomy for Sustainable Development. 2018;**38**(1)

[95] Phillips C et al. Physical feasibility of biochar production and utilization at a farm-scale: A case-study in non-irrigated seed production. Biomass and Bioenergy. 2018;**108**:244-251

[96] Raut M, Basu P, Acharya B. The effect of torrefaction pre-treatment on the gasification of biomass. International Journal of Renewable Energy & Biofuels. 2016:1-14

[97] Gaunt J, Lehmann J. Energy balance and emissions associated with biochar sequestration and pyrolysis bioenergy production. Environmental Science & Technology. 2008;**42**(11):4152-4158

[98] Bridgwater A. Fast pyrolysis processes for biomass. Renewable and Sustainable Energy Reviews. 2000;**4**(1):1-73

[99] Baker T, Bartle J, Dickson R, Polglase P, Schuck S. Prospects for bioenergy from short rotation crops in Australia. In: Third Meeting of IEA Bioenergy, Task, 1-15 1999. 1999

[100] Laird D. The charcoal vision: A win-win-win scenario for simultaneously producing bioenergy, permanently sequestering carbon, while improving soil and water quality. Agronomy Journal. 2008;**100**(1):178-181

[101] Lehmann J, Gaunt J, Rondon M. Bio-char sequestration in terrestrial ecosystems—A review. Mitigation and Adaptation Strategies for Global Change. 2006;**11**(2):403-427

[102] Hagemann N, Spokas K, Schmidt H-P, Kägi R, Böhler MA, Bucheli TD. Activated carbon, biochar and charcoal: Linkages and synergies across pyrogenic carbon's ABCs. Water. 2018;**10**:182, DOI: 10.3390/w10020182

[103] Demirbas A. Combustion characteristics of different biomass fuels. Progress in Energy and Combustion Science. 2004;**30**(2):219-230

[104] Food and Agriculture Organisation. World Food Security: A Reappraisal of the Concepts and Approaches. Rome: Food and Agricultural Organisation of the United Nations; 1983

[105] Bhattacharya S, Albina D, Abdul Salam P. Emission factors of wood and charcoal-fired cookstoves. Biomass and Bioenergy. 2002;**23**(6):453-469

[106] Pooe D, Mathu K. The South African coal mining industry: A need for a more efficient and collaborative supply chain. Journal of Transport and Supply Chain Management. 2011;**5**(1)

[107] Demirbaş A. Sustainable cofiring of biomass with coal. Energy Conversion and Management. 2003;**44**(9):1465-1479

[108] Temimi M, Camps J, Laquerbe M. Valorization of fly ash in the cold stabilization of clay materials. Resources, Conservation and Recycling. 1995;**15**(3-4):219-234

[109] Cho H, Oh D, Kim K. A study on removal characteristics of heavy metals from aqueous solution by fly ash. Journal of Hazardous Materials. 2005;**127**(1-3): 187-195

[110] Abdel-Raouf N, Al-Homaidan AA, Ibraheem IB. Microalgae and wastewater treatment. Saudi Journal of Biological Sciences. 2012;**19**(3):257-275. DOI: 10.1016/j.sjbs.2012.04.005

[111] EPA United States. Best Practices for Solid Waste Management: A Guide for Decision-Makers in Developing Countries. Environmental Protection Agency, EPA 530-R-20-002, 2020. [Online]. Available from: https://www.epa.gov/sites/production/files/2020-10/documents/master_swmg_10-20-20_0.pdf [Accessed: May 21, 2021]

[112] McAllister J. Factors Influencing Solid-Waste Management in the Developing World. Master of Science, All Graduate Plan B and other Reports. Logan, Utah: Utah State University; 2015. p. 528. [Online]. Available from: https://digitalcommons.usu.edu/gradreports/528

[113] Ozeler D, Yetis U, Demirer GN. Life cycle assesment of municipal solid waste management methods: Ankara case study. Environment International. 2006;**32**(3):404-411. DOI: 10.1016/j.envint.2005.10.002

[114] Yadav P, Samadder SR. Life cycle assessment of solid waste management options: A review. Recent Research in Science and Technology. 2014;**6**(1): 113-116

[115] Allesch A, Brunner PH. Material flow analysis as a tool to improve waste management systems: The case of Austria. Environmental Science & Technology. 2017;**51**(1):540-551. DOI: 10.1021/acs.est.6b04204

[116] Turner DA, Williams ID, Kemp S. Combined material flow analysis and life cycle assessment as a support tool for solid waste management decision making. Journal of Cleaner Production [Online]. 2016;**129**:234-248. DOI: 10.1016/j.jclepro.2016.04.077

[117] Thushari I, Vicheanteab J, Janjaroen D. Material flow analysis and life cycle assessment of solid waste management in urban green areas, Thailand, (in English). Sustainable Environment Research. 2020;**30**(1). DOI: 10.1186/s42834-020-00057-5

Chapter 4

Role of Microbial and Organic Amendments for the Enrichment of Methane Production in Bioreactor

Sharda Dhadse and Shanta Satyanarayan

Abstract

Studies were carried out on lab-scale levels for biogas production using two different wastewaters, that is, herbal pharmaceutical wastewater and food processing wastewater. A total of eight methane bacteria were isolated from cattle dung and mass culturing was carried out to study their feasibility in biogas escalation. Optimization of methane bacteria that could increase biogas production was identified. Among the methane bacteria, two species *Bacillus* sk1 and *Bacillus* sk2 were found to enhance the biogas production to a maximum level. Gas analysis showed CH_4 content of 63% in the case of food processing wastewater and around 67% with herbal pharmaceutical wastewater. *Bacillus* sk1 was found to be more suitable for both wastewater and biogas production and was found to be 46.4% in food processing wastewater and 43.3% in herbal pharmaceutical wastewater. Amendment of *Bacillus* sk2 in food processing wastewater produces 39.7% and 30.3% of biogas in herbal pharmaceutical wastewater was observed. Enzyme Bacillidine™ (P-COG-concentrate aqueous base) was also tried but results were not very encouraging. Comparative studies on both the wastewater have been discussed in detail in this article.

Keywords: anaerobic digestion, herbal pharmaceutical wastewater, food processing wastewater, methanogenesis, *Bacillus* sk1, *Bacillus* sk2

1. Introduction

Due to industrialization and excessive exploitation of natural resources, as well as the population explosion at the global level affects the environment at large [1, 2]. Textile industries, municipal sewage, dairy waste, pharmaceutical industries, swine, and aquaculture sectors release wastewater on a regular basis [3, 4]. Wastewater contains a variety of unfavorable chemical components and microbes that show short- and long-term environmental and human health implications [5, 6]. Untreated wastewater if utilized directly for irrigation may cause undesirable implications in the environment and groundwater [7]. The challenges with wastewater treatment include high energy consumption and laborious work [8]. In recent decades, a new goal has gained attraction on resource recovery from wastewater paired with its treatment technologies.

Regarding the crisis of energy use, the widespread usage of fossils fuels may deplete in the next 50 years [9]. So, to cope with the future energy demand, it is critical to seek innovative renewable energy sources [10, 11]. Other technologies for harnessing renewable energy sources, such as solar, wind, hydraulic, and geothermal energy, have been created [12]. All of these technologies have been developed very well and are commercially accessible to meet rising energy demand to some extent. Using wastewater as a source of energy can also help to relieve pressure on other technologies. Fresh microbial biomass or residual biomass after lipid extraction can be used directly for bioenergy generation using dark fermentation (biohydrogen production), fermentation (bio alcohols), and anaerobic digestion (methane) [13–15].

Presently, the global pharmaceutical sector has been quickly expanding and contributing to great economic development. But on the other hand, it is generating significant environmental degradation by releasing effluents [16]. Pharmaceutical manufacturing based on chemicals employs a number of chemical processes that result in complicated effluent with high salt content and poor nutritional value that is difficult to biodegrade. The wastewater generated by various pharmaceutical businesses is not uniform, and the composition of each type of wastewater is impacted by the techniques used. Antibiotics, steroids, reproductive hormones, analgesics, beta-lactamides, antidepressants, detergents, as well as unspent solvent and heavy metals make up the majority of the substance [17]. The treatment of wastewater includes several processes that are usually expensive. Anaerobic digestion (AD) is a low-cost technique for treating organic wastes while simultaneously recovering energy in the form of methane [18, 19]. The efficiency of anaerobic digestion is determined by the cooperation of numerous microorganisms that conduct hydrolysis, fermentation, and methanogenesis [20, 21]. Anaerobic digestion (AD) is a waste-to-biomethane conversion technology that has been utilized to transform sewage sludge, agricultural/livestock residues, food wastes, and other organic waste streams into biomethane [22]. Many research studies have been undertaken for the past 10 years to optimize the digestion benefits in terms of biogas production, environmental effect, and reduction of waste [23].

Therefore, looking at the present scenario of energy demand in a developing country like India, it is very much necessary to switch over to bio-methanation, as it is the ultimate environment-friendly and sustainable way of progressive development. Our study aimed to produce biogas from two types of wastewaters namely herbal medicinal wastewater and food processing wastewater using cow dung and isolated microbial species.

2. Materials and methods

Samples of influent cattle dung slurry were collected and analyzed for various physicochemical characteristics. Similarly, effluent samples were also collected from an anaerobic conventional digester for their characterization after 38 days. The physical and chemical parameters were determined according to the Standard Procedures [24]. Two different wastewaters viz., herbal pharmaceutical and food processing were also collected and its treatability studies were carried out. Herbal pharmaceutical wastewater was tried to treat with Vermi filters that produce a nutrient biosolid and vermiwash that promotes the growth of plants [25, 26].

3. Isolation of anaerobic bacteria

The potential bacterial species involved in biogas production were isolated from a fresh sample of cattle dung. During isolation, the sample was diluted as per

Digester	Concentration
First set	
FPW	200 ml seed + 800 ml food processing wastewater (FPW)
Bacillus sk2	200 ml seed + 800 ml FPW + isolate no. 2 (2 ml)
Bacillus sk1	200 ml seed + 800 ml FPW + isolate no. 4 (2 ml)
Enzyme bacillidine	200 ml seed + 800 ml FPW + enzyme *Bacillidine* (2 ml)
Second set	
HPW	200 ml seed + 800 ml herbal pharmaceutical wastewater (HPW)
Bacillus sk2	200 ml seed + 800 ml HPW + isolate no. 2 (2 ml)
Bacillus sk1	200 ml seed + 800 ml HPW + isolate no. 4 (2 ml)
Enzyme bacillidine	200 ml seed + 800 ml HPW + enzyme *Bacillidine* (2 ml)

Sterilized seeds were used.

Table 1.
Different digester mixture with seed.

requirement. The anaerobic agar medium was used for the selective isolation of the anaerobes. The spread-plate method was employed for bacterial-cell isolation. A total of 0.5 ml of diluted bacterial samples was spread onto an anaerobic agar plate. The plates were then incubated in an anaerobic jar provided with alkaline pyrogallol to make the environment free of molecular oxygen. Incubation was carried out for 72–96 h, at 37°C for bacteria proliferation.

Microscopic examination was carried out to study the presence of methane bacteria and their motility. A total of eight methanogenic cultures were isolated and detailed biochemical tests were carried out, such as the sugar fermentation test and IMViC test. Apart from these two tests Indole test, methyl red test, Voges Proskauer test, citrate utilization test, etc., were also evaluated. Enzyme like catalase test, oxidase test was performed to authenticate the presence of these enzymes. To confirm the methanogenic nature of the bacteria fluorescence test was carried out. Identification of the eight cultures of the methane bacteria was carried out by subjecting the isolates to scanning electron microscopy.

A total of eight isolates of bacterial species were obtained from the predigested cow dung slurry, which was inoculated for the comparative assessment of methane production individually by each species. The studies were conducted for two months, based on which the isolates 2 and 4 were contributing maximum. So, by taking these two isolates the experiment was designed by taking wastewaters from two different industries, namely the food processing industry and herbal pharmaceutical industry. Along with that, the enzyme *Bacillidine* was also taken for the experiment, which was isolated from the same material (pre-digested cow dung slurry). The details of different combinations taken for the experiment have given in **Table 1**.

4. Experimental set up

4.1 Part 1

Fresh cow dung slurry was prepared, for that, a total solid content was diluted in water to make a slurry of 1:1 ratio to get a total solid content of approximately 8.0% solids. Every digester was filled with 950 ml of diluted cattle dung slurry plus 50 ml of seed slurry from a working biogas plant. So, the working volume of digesters was kept to be 1 liter. A total of eight species of methane bacteria were isolated and they were subjected to mass culturing. Out of the nine digesters, one was kept as control with only cattle dung, while others were used for experimental purposes and named 1, 2, 3, 4, 5, 6, 7, and 8, respectively. These eight cultures individually were added (2 ml) to each of the above eight cattle dung digesters.

The digesters with maximum gas production were identified and further experimental work was initiated. Two bacterial species, that is, isolate 2 and isolate 4 resulted in maximum biogas production. The biogas production by individual species is given in **Figure 1**. DNA sequencings for 16s RNA studies of isolates 2 and 4 show that isolate 2 is *Bacillus* sk2 and isolate 4 is *Bacillus* sk1 (**Figures 2** and **3**).

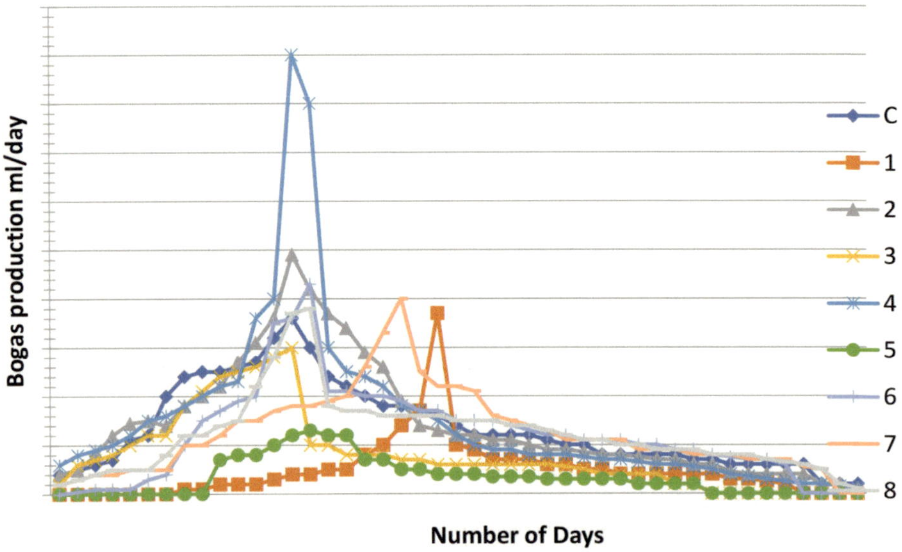

Figure 1.
*Daily biogas production by eight types of isolates. *C: control and 1, 2, 3, 4, 5, 6, 7, 8 are the isolates no 1, 2, 3, 4, 5, 6, 7, 8.*

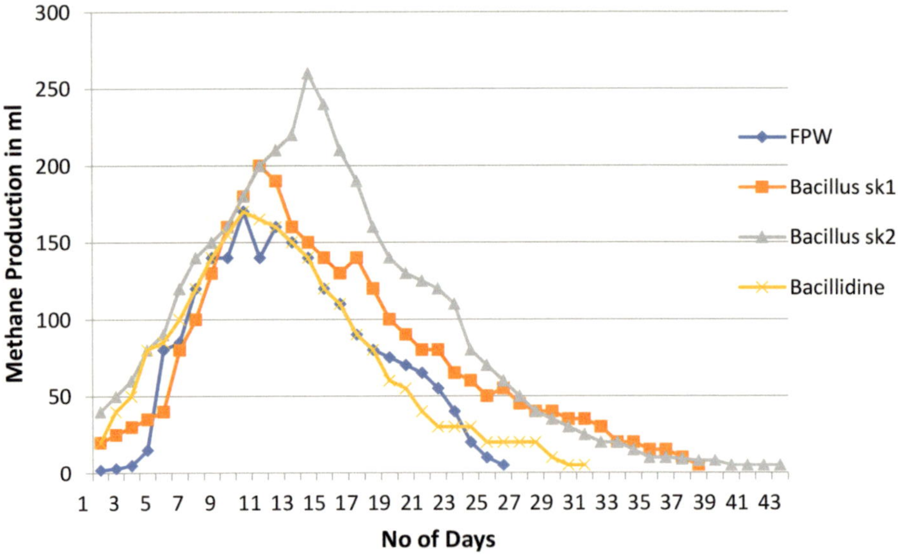

Figure 2.
*Daily biogas production in food processing wastewater with Bacillus sk1, Bacillus sk2, and enzyme Bacillidine. *FPW: food processing wastewater.*

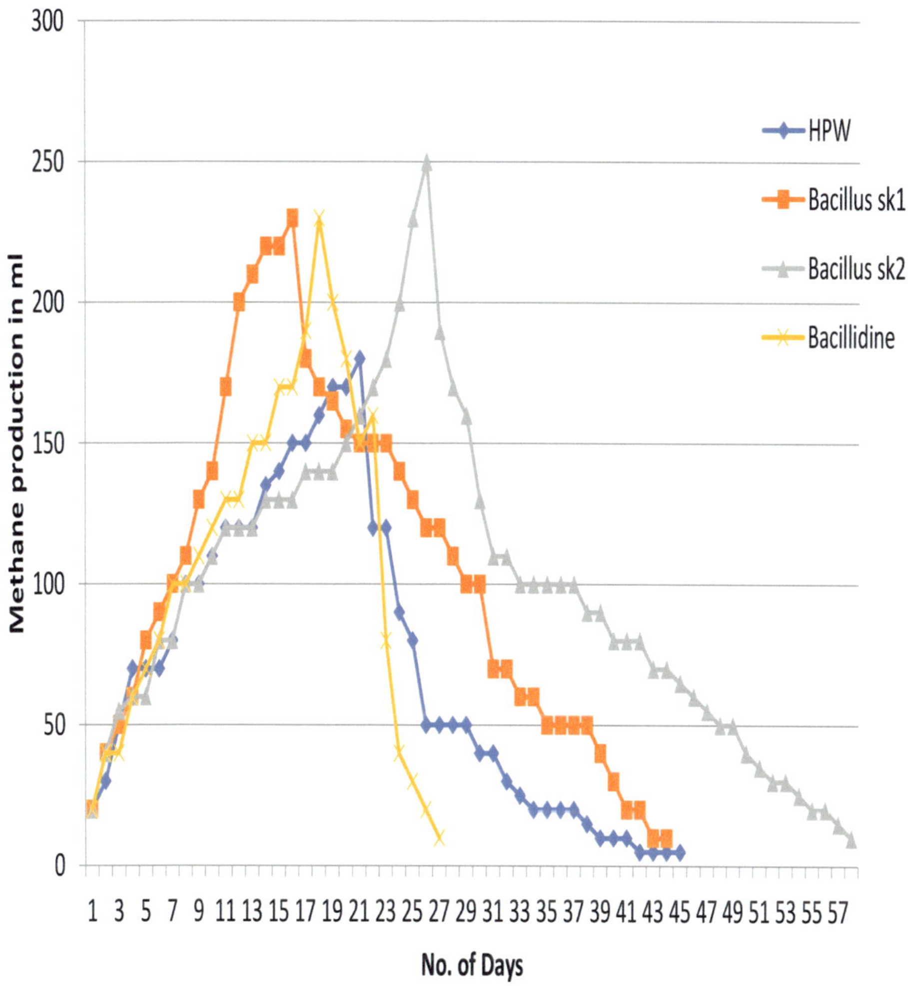

Figure 3.
*Daily biogas production by herbal pharmaceutical wastewater with Bacillus sk1, Bacillus sk2, and enzyme Bacillidine. *HPW: herbal pharmaceutical wastewater.*

4.2 Part 2

Eight digesters of 2-liter capacities were taken for two different wastewaters for different combinations with Bc (isolate 2) and Bt (isolate 4) and enzyme *Baccillidine* along with control as H and B. The biogas was collected in the inverted measuring cylinder of 2-liter capacity filled with 20% sodium sulfate [27]. Every day the digesters were manually shacked for 3–4 times a day and daily gas production was monitored. This situation was maintained for a period of 2 months till the gas production ceased completely. The digester where maximum gas production was obtained was selected for further experimental work using wastewaters as an organic amendment.

5. Results and discussion

There were many industrial wastewaters studied by anaerobic treatment by various authors. These wastewaters may have properties that accelerate the process of digestion. Some of the properties, such as temperature, pH, alkalinity, total ammonia nitrogen, volatile acids, total solids, total volatile solids, and phosphate, may also influence the enzymatic reaction of anaerobic treatment. In the present study, both the wastewaters were containing plant materials as organic matter. Therefore, it was easily degrading when processed anaerobically.

Karray et al. [28] utilized anaerobic digestion of green algae *Ulva rigida* with sugar industry influent and tried to increase biogas generation. The results showed that this combination helps to compost algae with anaerobic sludge and water yielded the optimal inoculum for producing biogas and feeding an anaerobic reactor, providing 408 mL of biogas. When sugar co-substrate was employed, a maximum methane generation yield of 114 mL g^{-1} was received with 75% methane. Biogas was produced anaerobically from wastewater of the Colombian palm oil mill industry by Nabarlatz et al. [29]. Using two distinct inoculums, anaerobic digestion tests were carried out in batch mode to assess the effects of pH and inoculum to substrate ratio on anaerobic digestion. The best-suited inoculum was determined to be a 1:1 v/v mixture of urban WWTP (wastewater treatment plant) anaerobic sludge/pig manure at a ratio of 2 g volatile solids (VS) inoculum/g VS substrate, which produced the largest amount of accumulated methane, attaining 2740 mL methane without neutralizing pH. Anaerobic digestion of brewery wastewater to enhance biogas production using UASB (upflow anaerobic sludge blanket) reactor was carried out by Enitan et al., [30]. Using a modified methane generation model, 1.46 L CH_4/g COD was generated. Similarly, Debowski et al. [31] investigated the anaerobic treatment of dairy wastewater in a multi-section horizontal flow reactor (HFAR) using microwave and ultrasonic generators. The study's findings in terms of wastewater treatment efficiency, biogas output, and economic analysis results demonstrated that the HFAR can compete with existing industrial technologies for food wastewater treatment [31]. Ounsaneha et al. [32], evaluated biogas generation during the digestion of municipal wastewater and food waste in semi-continuous and continuous operation with varying hydraulic retention times (HRTs). At 30 days of HRTs with a 10:90 ratio of municipal wastewater to food waste, methane outputs of 167.41 66.52 ml/g-Vs were observed in semi-continuous mode.

Initially, physicochemical parameters of seed slurry, herbal, and food processing wastewater were determined by the standard methods. The parameters that were determined included—pH, alkalinity (as $CaCO_3$ mg/l), total ammonia nitrogen (mg/l), volatile acids (mg/l as CH_3COOH), total solids (%), total volatile solids (%), and phosphate (mg/l). The pH of the seed slurry before the experiment was 6.91 indicating it was very less acidic and very close to the neutral pH. The alkalinity was found to be 1280 mg/l (as $CaCO_3$). Total ammonia nitrogen was 105.28 mg/l. Total acids were determined as CH_3COOH that was 312 mg/l. The percentages of total solids and total volatile solids were 5.50% and 3.60%, respectively. The data obtained are given in the tabular form in **Table 2**.

Sr. no.	Parameter	Seed slurry	HPW	FPW
1.	pH	6.91	6.07	6.78
2.	Alkalinity as $CaCO_3$ mg/l	1280	20.00	28.40
3.	Total ammonia nitrogen mg/l	105.28	24.64	140.00
4.	Volatile acids mg/l as CH_3COOH	312.00	60.00	384.00
5.	Total solids %	5.50	14.00	10.20
6.	Total volatile solids %	3.60	7.10	3.40
7.	Phosphate mg/l	—	2328.00	2067.00

Table 2.
Physicochemical characteristics of herbal and food processing wastewater.

The physicochemical characterization of herbal processing wastewater (HPW) and food processing wastewater (FPW) was done where the same parameters were determined as in the seed slurry. The pH of FPW was 6.07, slightly higher than that of HPW in which the pH was found to be 6.78. The total ammonia nitrogen in the HPW was 20 mg/l, while that in FPW was 28.40 mg/l. Volatile acid in HPW was 60 mg/l, whereas in FPW the amount was comparatively very higher at about 384 mg/l. Total solids in HPW and FPW were 14% and 10.20%, respectively. Total volatile solids in HPW were found to be 7.10% and in FPW it was 3.40%. The amount of phosphate in HPW and FPW was found to be 2328 mg/l and 2067 mg/l, respectively. All the data obtained by characterization of HPW and FPW is shown in tabular form in **Table 2**.

Two sets of different digester mixtures were prepared for the experiment. In the first set food processing waste (FPW) water was used, whereas in the second set herbal processing wastewater (HPW) was used to make the digester mixtures with seed slurry. Each set contained four different digester mixtures. One mixture was prepared by mixing 200 ml of seed slurry and 800 ml of wastewater only. Another mixture contained 200 ml seed slurry, 800 ml wastewater, and isolate no. 2, that is, *Bacillus cereus* (Bc). The third kind of mixture contained 200 ml seed slurry, 800 ml wastewater, and isolate no. 4, that is, *Bacillus thuringiensis* (Bt). In the fourth kind of mixture enzyme, *Bacillidine* was added to 200 ml seed slurry and 800 ml wastewater. In this mixture, isolates 2, 4, and enzyme (2 ml) were mixed in each reactor. Hence four kinds of mixtures were there and the total numbers of eight digester mixtures were prepared using FPW and HPW. The above-mentioned information is summarized in **Table 1**.

The sets prepared were used in the experiment and kept for 45 days for biogas production. After 45 days the physicochemical parameters of all eight mixtures were determined. The same parameters were determined that were found initially. The pH of the four FPW effluents was slightly acidic to almost neutral. The alkalinity was between the range of 350–580 mg/l. The total ammonia was found in between 142 and 168 mg/l. Volatile acids were in the range of 168–276 mg/l. Total solids in the four mixtures were found in the range of 8–9%. Total volatile solids were between 6 and 7.2%. The pH of the four HPW effluents was slightly acidic compared to those of FPW. All the mixtures were ranged from 6.38 to 6.72. The alkalinity was between the range of 236–450 mg/l. The total ammonia was found in between 32 and 70 mg/l. Volatile acids were in the range of 120–218 mg/l. Total solids in the four mixtures were found in the range of 8.4–8.9%. Total volatile solids were between 5.7 and 8%.

6. Selection of isolates

The second part of this experiment showed that isolate no. 2, 8, and control were continued up to the 45th day, while isolate no. 4 and 7 has stopped on the 42nd day. Isolate no. 1 and 6 were stopped on the 40th day and isolate no 3 and 5 were stopped on the 36th day. On the 45th-day total biogas production was found to be 6080 ml in control, whereas in isolate 1, 3, 5, 6, 7, and 8 it was 2115 ml, 3595 ml, 1515 ml, 5430 ml, 5555 ml, and 5445 ml, respectively. The reactor containing isolate no. 2 was able to produce 6470 ml and the reactor with isolate no. 4 was able to produce 6900 ml of biogas, which was subsequently higher than isolate no. 2. So, it proves that isolate no 2 produces 6.4% and isolate no. 4 produces 13.5% more biogas as compared to control. **Figure 1** shows that isolate no. 4 was having the highest peak on the 13th day with production of 900 ml of biogas. Therefore, isolate no. 2 and 4 were selected for further studies.

A fluorescence test was conducted for the identification of methanogenic bacteria having the F_{420} coenzyme, which depicts blue-green fluorescence by methanogenic bacteria and was easily differentiated from the white-yellow fluorescence observed in non-methanogenic bacteria. The isolate no. 2, 6, 7, and 8 indicated the blue-green fluorescence in ultraviolet light depicting the presence of methanogenic bacteria, whereas isolate no. 1, 3, 4, and 5 indicated the negative fluorescence activity.

7. Efficient isolate along with organic additive

The experiment conducted to prove the efficiency of isolats no. 2 and 4 along with the organic additives like food processing wastewater and herbal pharmaceutical wastewater were continued till the complete anaerobic digestion took place. The digester mixtures were prepared as given in **Table 1**. After completing anaerobic digestion, the physicochemical analysis shows that they are well within the limits as per standards. The physicochemical characteristic of seed slurry and wastewater is given in **Table 2**. Initial characteristics of seed slurry, herbal, and food processing wastewater were determined by the standard methods. The parameters included were pH, alkalinity (as $CaCO_3$ mg/l), total ammonia nitrogen (mg/l), volatile acids (mg/l as CH_3COOH), total solids (%), total volatile solids (%), and phosphate (mg/l). The pH of the seed slurry before the experiment was 6.91 indicating it was very less acidic and very near to the neutral pH. The characterization of herbal processing wastewater (HPW) and food processing wastewater (FPW) showed the pH of FPW was 6.07, which was slightly higher than the HPW in which the pH was found to be 6.78.

Two sets of different digester mixtures were prepared for the experiment. In the first set food processing waste (FPW) water was used, whereas in the second set herbal processing waste (HPW) water was used to make the digester mixtures with seed slurry. Each set contained four different digester mixtures. The mixture was prepared by mixing 200 ml of seed slurry and 800 ml of wastewater with an inoculum of isolate, which was given in **Table 1**.

The digesters containing food industrial wastewaters were continued for 43 days, while digesters of herbal pharmaceutical wastewaters were continued for 58 days. The physicochemical characteristics of completely digested effluents were given in **Table 3**. Results indicated pH in the range of 6.58–7.08 and volatile acid to alkalinity ratio well below 0.8 indicated good buffering. Total ammonia nitrogen was well within the limits indicating the efficient working of reactions takes place. In no instances, there was any alarming increase in either volatile acid or total ammonia nitrogen shows that the system was well-balanced methane activity. This was due to the presence of higher organic content in the wastewaters.

The characteristics of effluents after 45 days of food processing wastewater and herbal pharmaceutical wastewaters were given in **Table 3**. The pH of the four FPW effluents was slightly acidic to almost neutral. The alkalinity was between the range of 350–580 mg/l, with total ammonia as 142–168 mg/l. Whereas, volatile acids were in the range of 168–276 mg/l. Total solids in the four mixtures were found in the range of 8–9%. With total volatile solids in between 6 and 7.2%.

The pH of the four HPW effluents was slightly acidic compared to those of FPW. All the mixtures were ranged from 6.38 to 6.72. The alkalinity was between the ranges of 236–450 mg/l. The total ammonia was found in between 32 and 70 mg/l. Volatile acids were in the range of 120–218 mg/l. Total solids in the four mixtures were found in the range of 8.4–8.9%. Total volatile solids were between 5.7 and 8%. The effluent of FPW and HPW showed minimized total solids after exhausting the bioreactors.

Sr. no.	Parameter	HPW effluents				FPW effluents			
		HPW	*Bacillus* sk2	*Bacillus* sk1	Enzyme bacillidine	FPW	*Bacillus* sk2	*Bacillus* sk1	Enzyme bacillidine
1.	pH	6.72	6.58	6.66	6.38	7.07	6.93	6.84	7.08
2.	Alkalinity as $CaCO_3$ mg/l	450.80	290.80	320.80	236.00	350.20	580.80	386.40	520.00
3.	Total ammonia nitrogen mg/l	46.20	32.00	70.56	56.28	159.60	168.00	147.84	142.80
4.	Volatile acids mg/l as CH_3COOH	218.40	156.00	168.00	120.00	180.00	276.00	168.00	240.00
5.	Total solids %	8.80	8.50	8.40	8.90	8.0	8.50	9.00	8.80
6.	Total volatile solids %	8.00	6.00	7.10	5.70	7.2	6.04	6.40	6.80

Table 3.
Characteristics of effluents with herbal pharmaceutical wastewater (HPW) and food processing wastewater (FPW).

8. Methane production

The microbial activity may get affected by some of the factors in an anaerobic digester. The design of the reactor, temperature, pH, C:N ratio, and wastewater characteristics along with the composition of complete seed material. Some of the methanogens belonging to the order viz., methanosarcinales, methanosarcinaceae, and methanosaetaceae may often be detected in accelerating methanogenicity.

Ho and Sung [33] investigated methanogenic activity in anaerobic membrane bioreactors (AnMBRs) used to treat synthetic municipal wastewater. The methanogenic activity profiles of suspended and attached sludge in AnMBRs treating synthetic municipal wastewater at 25 and 15°C were investigated using the specific methanogenic activity (SMA) assay. On day 1, AnMBR 1's methanogenic activity was 51.8 ml CH4/g VSS d, but by day 75, it had grown by 27% to 65.7 ml CH4/g VSS d. The methanogenic activity of AnMBR 2 sludge, on the other hand, was lower than that of AnMBR 1. Silva et al. [34] looked at the effects of pharmaceuticals like Ciprofloxacin (CIP), Diclofenac (DCF), Ibuprofen (IBP), and 17α ethinylestradiol (EE2) on the activity of acetogens and methanogens in anaerobic communities. The majority of these compounds end up in wastewater treatment plants. The specific methanogenic activity was unaffected at doses of 0.01–0.1 mg/L. Acetogenic bacteria were sensitive to CIP concentrations more than 1 mg/L, whereas DCF and EE2 toxicity was only identified at concentrations greater than 10 mg/L, and IBP had no effect at any concentration. Acetoclastic methanogens were more sensitive to these micropollutants, being affected by all of the pharmaceutical chemicals tested, but to varying degrees. When compared to acetoclasts and acetogens, hydrogenotrophic methanogens were unaffected by any concentration, showing that they are less sensitive to these chemicals. CIP had the greatest impact on microbial communities, followed by EE2, DCF, and IBP, but the responses of the various microbial species differed [34]. The co-digestion of mixed sludge from wastewater treatment plants and the organic fraction of municipal solid trash were explored by Keucken et al. [35]. When co-digesting mixed sludge with organic fraction at a 1:1 ratio, based on the volatile solids (VS) concentration, the results reveal rapid adaptability of the process and an increase in biomethane output of 20–40%. The microbial community is also affected by the introduction of organic fractions. The methanogenic activity grows and adapts to acetate decomposition under 50% co-substrate and constant loading circumstances (1 kg VS/m^3/d), while the community in the reference reactor, which does not have a co-substrate, remains unaffected. The methanogenic activity in both reactors increases when the load is increased (2 kg VS/m^3/d), while the composition of the methanogenic population in the reference reactor remains unchanged [35].

Isolate no. 4 was found to be more suitable for herbal wastewater than food processing wastewater because biogas production was almost double in the case of herbal pharmaceutical wastewater. Enzyme Bacillidine™ (P-COG-concentrate aqueous base) was also tried but results were not very encouraging. In the case of herbal pharmaceutical wastewater also the increase in biogas production was very significant with isolate 2 with total gas production of 3085 ml as compared to 2068 ml in the case of food processing wastewater.

The bioreactor with food processing wastewater and only sterilized seed was able to produce the biogas up to the 18th day, which was 2090 ml. However, the bioreactor with Bc was able to produce the biogas for 38 days with 2920 ml, which was almost 39.7% higher than the control, while the bioreactor with Bt was able to produce 3895 ml of biogas in 43 days, which was 86.4% higher than the control. In the case of enzyme Baccilidine, the biogas production was observed to be 2320 ml in 39 days, which was only 11.0% higher than the control. **Figure 2** shows that food processing wastewater produces 2090 ml of biogas in 43 days, but bioreactor amended with the culture of Bc produces 2920 ml of biogas that was 39.7% more and Bt amended bioreactor producing 3895 ml of biogas, which was 46.4% more than control. In the case of enzyme *Baccillidine*, only 2320 ml of biogas was produced, which was only 10.2% higher than the control.

Similar results were observed in the case of herbal pharmaceutical wastewaters also. **Figure 3** shows that the bioreactor containing only wastewater and seed was able to produce 3205 ml of biogas in 45 days, whereas the bioreactor containing isolate 2 was able to produce 4600 ml of biogas in 44 days that was 43.5% higher than the control and the bioreactor containing isolate 4 was able to produce 5650 ml of biogas in 58 days, which was 76.3% higher than the control.

Enzyme *baccilidine* was able to produce only 2930 ml of biogas in 27 days, which was 8.6% lesser than the control. Hence it has been proved that the Bt contributes more than Bc for the biogas production, while enzyme *baccilidine* attenuates the biogas production in overall processes. In the case of enzyme *Baccillidine,* it produced only 2750 ml (14.0% less than control) of biogas.

9. Conclusion

Looking at the present scenario of the energy crisis and the environmental damage that occurs due to the use of nonrenewable sources of fossil fuel, it is the need of an hour to switch over to the use of renewable sources of energy. In the present studies, the herbal pharmaceutical wastewater and food industry wastewater were rich in organic content, which promotes the anaerobic biodegradability with a maximum production of methane by inoculating the specific bacteria isolated from the cow dung. Only cow dung seed is not able to produce more biogas as compared to isolated microbes. *Baccilus* sk1 (Bt, isolate no. 4) was highly capable of methane production rather than *Bacillus* sk2I (Bc, isolate 2). Therefore, the culture of *Bacillus* sk1 became the best enhancer of biogas production. Such inoculums can be cultured on large scale and may be utilized for future energy generation.

Acknowledgements

The author is grateful to University Grant Commission (UGC) for providing the fellowship.

Author details

Sharda Dhadse* and Shanta Satyanarayan
CSIR-National Environmental Engineering Research Institute, Nehru Marg, Nagpur, India

*Address all correspondence to: sn_dhadse@neeri.res.in

References

[1] Bhatia SK, Joo H-S, Yang Y-H. Biowaste-to-bioenergy using biological methods—A mini-review. Energy Conversion and Management. 2018;**177**:640-660. DOI: 10.1016/j.enconman.2018.09.090

[2] Manaf ISA, Embong NH, Khazaai SNM, Rahim MHA, Yusoff MM, Lee KT, et al. A review for key challenges of the development of biodiesel industry. Energy Conversion and Management. 2019;**185**:508-517. DOI: 10.1016/j.enconman.2019.02.019

[3] Kadir WNA, Lam MK, Uemura Y, Lim JW, Lee KT. Harvesting and pre-treatment of microalgae cultivated in wastewater for biodiesel production: A review. Energy Conversion and Management. 2018;**171**:1416-1429. DOI: 10.1016/j.enconman.2018.06.074

[4] Lellis B, Fávaro-Polonio CZ, Pamphile JA, Polonio JC. Effects of textile dyes on health and the environment and bioremediation potential of living organisms. Biotechnology Research and Innovation. 2019;**3**:275-290. DOI: 10.1016/j.biori.2019.09.001

[5] Bhatia SK, Mehariya S, Bhatia RK, Kumar M, Pugazhendhi A, et al. Wastewater based microalgal biorefinery for bioenergy production: Progress and challenges. Science of the Total Environment. 2021;**751**:141599. DOI: 10.1016/j.scitotenv.2020.141599

[6] Fung FM, Su M, Feng H-T, Li SFY. Extraction, separation and characterization of endotoxins in water samples using solid phase extraction and capillary electrophoresis-laser induced fluorescence. Scientific Reports. 2017;7:10774. Available from: https://www.nature.com/articles/s41598-017-11232-x

[7] Libutti A, Gatta G, Gagliardi A, Vergine P, Pollice A, Beneduce L, et al. Agro-industrial wastewater reuse for irrigation of a vegetable crop succession under Mediterranean conditions. Agricultural Water Management. 2018;**196**:1-14. DOI: 10.1016/j.agwat.2017.10.015

[8] Sparn B, Hunsberger R. Opportunities and Challenges for Water and Wastewater Industries to Provide Exchangeable Services. Golden, CO (United States): National Renewable Energy Lab. (NREL); 2015

[9] Kumar R, Ghosh AK, Pal P. Synergy of biofuel production with waste remediation along with value-added co-products recovery through microalgae cultivation: A review of membrane-integrated green approach. Science of the Total Environment. 2020;**698**:134169. DOI: 10.1016/j.scitotenv.2019.134169

[10] Bhatia SK, Kim S-H, Yoon J-J, Yang Y-H. Current status and strategies for second generation biofuel production using microbial systems. Energy Conversion and Management. 2017;**148**:1142-1156. DOI: 10.1016/j.enconman.2017.06.073

[11] Mehariya S, Patel AK, Obulisamy PK, Punniyakotti E, Wong JWC. Co-digestion of food waste and sewage sludge for methane production: Current status and perspective. Bioresource Technology. 2018;**265**:519-531. DOI: 10.1016/j.biortech.2018.04.030

[12] Qazi A, Hussain F, Rahim NA, Hardaker G, Alghazzawi D, Shaban K, et al. Towards sustainable energy: A systematic review of renewable energy sources, technologies, and public opinions. IEEE Access. 2019;7:63837-63851. DOI: 10.1109/ACCESS.2019.2906402

[13] Sanchez Rizza L, Coronel CD, Sanz Smachetti ME, Do Nascimento M,

Curatti L. A semi-closed loop microalgal biomass production-platform for ethanol from renewable sources of nitrogen and phosphorous. Journal of Cleaner Production. 2019;**219**:217-224. DOI: 10.1016/j.jclepro.2019.01.311

[14] Solé-Bundó M, Garfí M, Ferrer I. Pretreatment and co-digestion of microalgae, sludge and fat oil and grease (FOG) from microalgae-based wastewater treatment plants. Bioresource Technology. 2020;**298**:122563. DOI: 10.1016/j.biortech.2019.122563

[15] Wirth R, Lakatos G, Böjti T, Maróti G, Bagi Z, Rákhely G, et al. Anaerobic gaseous biofuel production using microalgal biomass—A review. Anaerobe. 2018;**52**:1-8. DOI: 10.1016/j.anaerobe.2018.05.008

[16] Guo Y, Qi P, Liu Y. A review on advanced treatment of pharmaceutical wastewater. In: IOP Conference Series: Earth and Environmental Science. China: IOP Publishing; 2017. p. 012025. DOI: 10.1088/1755-1315/63/1/012025

[17] Rana RS, Singh P, Kandari V, Singh R, Dobhal R, Gupta S. A review on characterization and bioremediation of pharmaceutical industries' wastewater: An Indian perspective. Applied Water Science. 2017;**7**:1-12. DOI: 10.1007/s13201-014-0225-3

[18] Yu N, Guo B, Zhang Y, Zhang L, Zhou Y, Liu Y. Different micro-aeration rates facilitate production of different end-products from source-diverted blackwater. Water Research. 2020;**117**:115783. DOI: 10.1016/j.watres

[19] Zhang Q, Zhang L, Guo B, Liu Y. Mesophiles outperform thermophiles in the anaerobic digestion of blackwater with kitchen residuals: Insights into process limitations. Waste Management. 2020;**105**:279-288

[20] Sharda D, Kankal NC, Bharti K. Study of diverse methanogenic and non-methanogenic bacteria used for the enhancement of biogas production. International Journal of Life Sciences Biotechnology and Pharma Research. 2012;**1**(2):176-191

[21] Gao M, Guo B, Zhang L, Zhang Y, Liu Y. Microbial community dynamics in anaerobic digesters treating conventional and vacuum toilet flushed blackwater. Water Research. 2019;**160**: 249-258

[22] Jiang J, Li L, Cui M, Zhang F, Liu Y, Liu Y, et al. Anaerobic digestion of kitchen waste: The effects of source, concentration, and temperature. Biochemical Engineering Journal. 2018;**135**:91-97. DOI: 10.1016/j.bej.2018.04.004

[23] Mata-Alvarez J, Dosta J, Mace S, Astals S. Codigestion of solid wastes: A review of its uses and perspectives including modeling. Critical Reviews in Biotechnology. 2011;**31**(2):99-111. DOI: 10.3109/07388551.2010.525496

[24] APHA, AWWA, WEF. Standard Methods for the Examination of Water and Wastewater. 20th ed. Washington DC: EUA; 2000

[25] Sharda D, Chaudhari PR, Shanta S, Wate SR. Vermitreatment of pharmaceutical wastewaters and nutrient bioassay of treated effluents for reuse as irrigation water. American Journal of Engineering Research (AJER). 2014;**3**(8):113-123

[26] Sharda D, Shanta S, Chaudhari PR, Wate SR. Vermifilters: A tool for sustainable aerobic biological treatment of herbal pharmaceutical wastewater. Water Science and Technology: A Journal of the International Association on Water Pollution Research. 2010;**61**(9):2375-2380

[27] Mamun MRI, Torii S. Enhancement of methane concentration by removing contaminants from biogas mixtures

using combined method of absorption and adsorption. International Journal of Chemical Engineering. 2017;**2017**:Article ID 7906859. DOI: 10.1155/2017/7906859

[28] Karray R, Karray F, Sayadi S. Anaerobic co-digestion of Tunisian green macroalgae Ulva rigida with sugar industry wastewater for biogas and methane production enhancement. Waste Management. 2017;**61**:171-178. DOI: 10.1016/j.wasman.2016.11.042

[29] Nabarlatz DA, Arenas-Beltrán LP, Niño-Bonilla DA. Biogas production by anaerobic digestion of wastewater from palm oil mill industry. CTyF—Ciencia, Tecnologia y Futuro. 2013;**5**:73-84. DOI: 10.29047/01225383.58

[30] Enitan AM, Adeyemo J, Bux F. Anaerobic digestion model to enhance treatment of brewery wastewater for biogas production using UASB reactor. Environmental Modeling and Assessment. 2015;**20**:673-685. DOI: 10.1007/s10666-015-9457-3

[31] Debowski M, Zielinski M, Kazimierowicz J. Evaluation of anaerobic digestion of dairy wastewater in an innovative multi-section horizontal flow reactor. Energies. 2020;**13**(9):115783. DOI: 10.3390/en13092392

[32] Ounsaneha W, Rattanapan C, Rakkamon T. Biogas production by co-digestion of municipal wastewater and food waste: Performance in semi-continuous and continuous operation. Water Environment Research. 2021;**93**:306-315. DOI: 10.1002/wer.1413

[33] Ho J, Sung S. Methanogenic activities in anaerobic membrane bioreactors (AnMBR) treating synthetic municipal wastewater. Bioresource Technology. 2010;**101**:2191-2196. DOI: 10.1016/j.biortech.2009.11.042

[34] Silva AR, Gomes JC, Pereira L. Ciprofloxacin, diclofenac, ibuprofen and 17α-ethinylestradiol differentially affect the activity of acetogens and methanogens in anaerobic communities. Ecotoxicology. 2020;**29**:866-875. DOI: 10.1007/s10646-020-02256-7

[35] Keucken A, Habagil M, Arnell M. Anaerobic co-digestion of sludge and organic food waste-performance, inhibition, and impact on the microbial community. Energies. 2018;**11**(9):2325. DOI: 10.3390/en11092325

Chapter 5

Global Fertilizer Contributions from Specific Biogas Coproduct

Sammy N. Aso, Simeon C. Achinewhu and Madu O. Iwe

Abstract

The impact of Haber-Bosch process on modern agriculture is prodigious. Haber-Bosch process led to invention of chemical fertilizers that powered green revolution, minimized food scarcity, and improved human and animal nutrition. Haber–Bosch process facilitated agricultural productivity in many parts of the world, with up to 60% of crop yield increase attributed solely to nitrogen fertilizer. However, Haber-Bosch fertilizers are expensive, and their poor use efficiency exerts adverse external consequences. In European Union for example, the annual damage of up to € 320 (US$ 372.495) billion associated with chemical fertilizers outweighs their direct benefit to farmers, in terms of crops grown, of up to € 80 (US $ 93.124) billion. A substitute for chemical fertilizers is therefore needed. In this chapter, external costs of chemical fertilizers are highlighted. The capability of liquid fraction of cassava peeling residue digestate to supplant and mitigate pecuniary costs of chemical fertilizers required for production of cassava root is also analyzed and presented. Results indicate that about 25% of fund used to purchase chemical fertilizers required for cassava root production could be saved with the use of liquid fraction of cassava peeling residue digestate. The pecuniary value is estimated at US$ 0.141 (≈ € 0.121) billion for the 2019 global cassava root output. This saving excludes external costs associated with Haber-Bosch fertilizers such as ammonia air pollution, eutrophication, greenhouse gasses emissions, and contamination of potable water supply reserves. Consequently, liquid fraction digestate could reduce the cost of cassava root production, as well as minimize adverse health and environmental consequences attributed to chemical fertilizers.

Keywords: anaerobic digestion, biogas, cassava peeling residue (CPR), chemical fertilizer, circular economy, cost savings, digestate, eutrophication, Haber-Bosch process

1. Introduction

The impact of Haber-Bosch process on modern agriculture may not be overemphasized. It led to the invention of inorganic fertilizers that powered global green revolution, minimized food scarcity, and improved human and animal nutrition. In his noble lecture, Fritz Haber (The 1918 noble laureate for chemistry; for the Haber-Bosch process) alluded that his impetuses for creation of ammonia from the elements were to meet increasing human food requirements, and replenish soil nitrogen extracted by harvested crops when he concluded: "*Let it suffice that in the meantime improved nitrogen fertilization of the soil brings new nutritive riches to mankind and that the chemical industry comes to the aid of the farmer who, in the good earth, changes stones into bread*" [1]. Haber–Bosch process has facilitated agricultural productivity in many parts of the world, with up to 60% of crop yield increase attributed solely to nitrogen fertilizer [2]. It has been estimated that between 1908 and 2008, Haber–Bosch nitrogen enabled the number of humans sustained per hectare of arable land to increase from 1.9 to 4.3 persons [3]. However, poor nitrogen use efficiency (NUE) of the same fertilizer that laid the golden benefits has deposited unintended adverse consequences to environmental systems [4]. Impacts of poor NUE may manifest at local, regional, and global scales [5], thereby placing air, soil, and water quality and safety, as well as human and animal health in jeopardy. Environmental and ecosystem services disruptions due to fertilizer use in agriculture have been reported worldwide. These include impairments of eco-diversity, recreational use of freshwaters, lakefront property values, and drinking water supply sources [6, 7]; loss of tourism benefits to coastal communities, [8, 9]; greenhouse gas (GHG) emissions and climate perturbation [10]; as well as air quality degradation [11].

Ammonia (NH_3) air pollution from animal husbandry, fertilizer production and application has also been documented and reported [12–14]. About 94% of NH_3 emissions in Italy emanate from agricultural operations [15], and in 2013 and 2018, agriculture contributed 93% of all ammonia emissions in the European Union [16, 17]. In the United States, agricultural runoff and drainage accounts for 89% of the total nitrogen inputs into the Mississippi River [18], contributing to hypoxic zone of the Gulf of Mexico [19]. In France, about 89% of residual nitrogen contamination of water resources and marine environments is attributed to mineral fertilizer and animal manure [8]. Similarly, nitrate contamination of surface and ground water is associated with agricultural use of fertilizers and manures [7, 8, 20–23]. Nitrous oxide (N_2O) is a greenhouse gas that contributes to stratospheric ozone shield depletion and climate change [10]. Nitrogen fertilizer and manure contribute 92% of all N_2O attributable to agriculture in the USA [24, 25]. In Italy and China, fertilizer accounts for about 68% of annual N_2O emissions [15, 26]. Chemical fertilizer and manure are major contributors to external costs such as eutrophication and acidification of ecosystems [21, 27–30]. Annually, up to € 320 (US$ 372.495) billion damage is associated with the use of nitrogen fertilizers in the European Union compared to direct economic benefit to farmers, in terms of crops grown, estimated at up to € 80 (US$ 93.124) billion [31]. Report currency, € 1.0 ≈ US$ 1.164 based on currency converter site: https://www1.oanda.com/currency/converter/as at Friday 22nd October 2021. Furthermore, inorganic fertilizers are not cheap, and may be used in large quantities. As at the second week of September 2021, the cost of 1 kg of nutrient fertilizer could range from ≈ US$ 0.375 for liquid nitrogen (as urea) to US$ 0.807 for dry phosphorus (as P_2O_5) [Ramsdell F&M Ltd. Brookings, SD USA]. In 2019, approximately 188.54×10^9 kg nutrient fertilizers (including 107.74×10^9 kg N, 43.41×10^9 kg P_2O_5, and 37.39×10^9 kg K_2O) were consumed in agricultural production globally [32].

Due to outlined adverse effects and financial exigencies of chemical fertilizers, a more sustainable, environmentally benign, and cost-effective fertilizer system is desired. Digestate in the context of circular economy could play a prominent role. In this chapter, cost implications of using liquid fraction (LF) of cassava peeling residue (CPR) digestate, to supplement chemical fertilizers required for cassava root production are analyzed and presented (**Figure 1**).

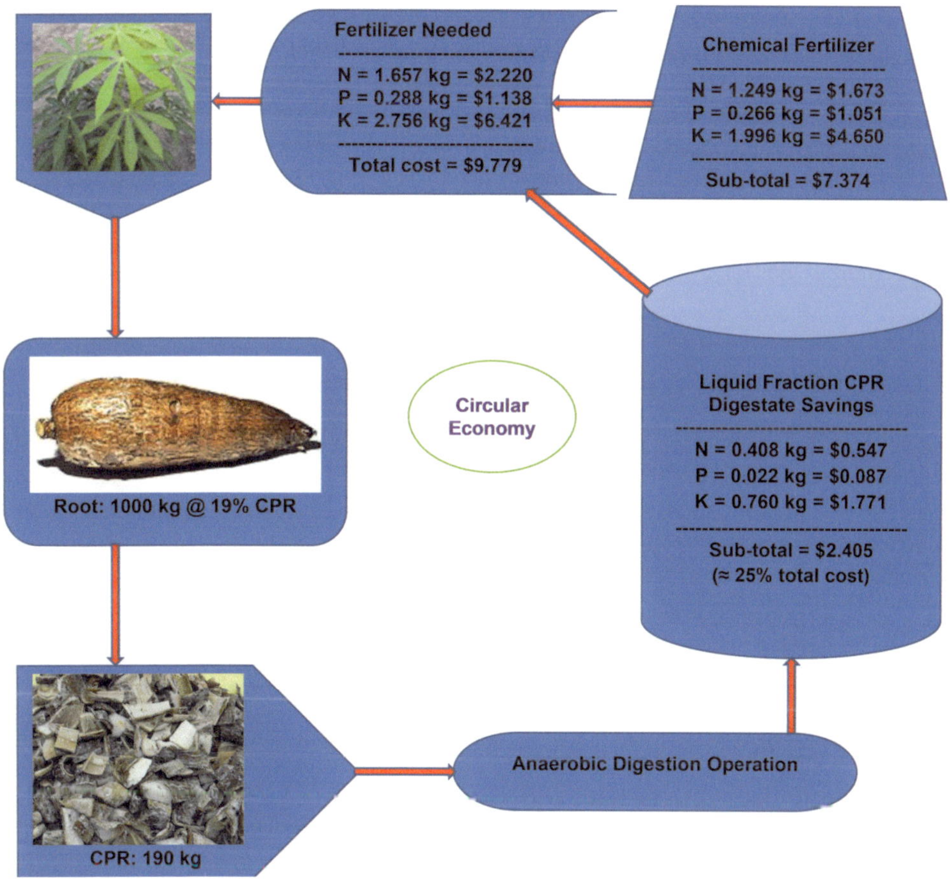

Figure 1.
Graphical representation of the objectives and summary of this chapter.

2. Anaerobic digestion and digestate for circular economy

Circular economy is a credible intervention tool to minimize GHG emissions, limit global warming and ecosystem degradation. The circular economic model maximizes material and product conservation; prudent consumption; eco-friendly biorefinery; recyclability and reusability; green- smart mobility and renewable energy; systems thinking, innovative business models and policies; wasteless design and zero waste cities, as well as generation of useful products out of waste [33–40]. Anaerobic digestion (AD) is a responsive technology that could rise to the occasion. In the context of biorefinery platform, sustainability, and circular economy, AD transforms organic matter to two major coproducts: biogas fuel and digestate [41]. Digestate has soil amendment and biofertilizer potentials.

Digestate enhances soil biological stability and enzymatic activities [42]; enriches microbial biomass [42, 43]; abates nutrients leaching and remediates metal contaminants [43–45]; conditions the soil and boosts plant nutrients, stimulates growth of beneficial microbes, improves buffering capacity, and physical properties such as texture, aeration, bulk density, hydraulic conductivity, and moisture retention capacity [46–49]. In comparison to chemical fertilizers, digestate biofertilizers offer better ecosystem services, values, and life cycle assessment accounting [50]; including lower energy consumption [51–53], lower ammonia air pollution [15], lower GHG emissions [53–55], better soil carbon sequestration [54, 56], reduced soil erosion [54, 57, 58], and increased biodiversity [59].

To exploit these benefits and advantages, various organic substrates have been used for digestate creation via the AD process. At least 120 items have been identified in published scientific literature [41], but CPR is not one of them. Indeed, there is scarcity of information on nutrient content, speciation, agronomic properties and values of LF of digestates from AD of single feedstocks in general [60]; and LF of digestate derived from AD of CPR as single feedstock in particular [41, 61].

3. Nutrient content of liquid fraction (LF) of cassava peeling residue (CPR) digestate

The only information on primary macronutrients (i.e., nitrogen (N), phosphorus (P), and potassium (K)) content of LF digestate of CPR as sole feedstock found in literature is presented in **Table 1**. For perspective, the values are compared with LF of digestates derived from other feedstocks in **Tables 2–4** respectively for N, P, and K. The values for each Table are presented in descending magnitude order. It can be seen that LF of CPR digestate is high in N and K, but low in P. Apart from livestock manure, most LF digestates with higher nutrient values are derived from AD of multiple feedstocks (**Tables 2–4**). Co-digestion of feedstocks may benefit from coactive effects.

S/N	Nutrient	Value [mg/L]
1	Total Kjeldahl nitrogen (N)	573
2	Total phosphorus (P)	31
3	Total Potassium (K)	1066

Table 1.
Macronutrients (N, P, K) content of liquid fraction of CPR digestate [41].

S/N	Feedstock	N Value [mg/L]	Reference
1	Cow manure & slurry (70%), maize silage (20%) and grass silage (10%)	5591	[62]
2	Dairy manure	4723	[63]
3	Cattle & pig slurries (main feedstocks), various food wastes (co-substrates)	4268–4507	[64]
4	Dairy cow slurry	2800–4500	[65]
5	Organic waste (Kitchen garbage, spoilt food, etc.)	3610–4120	[66]
6	Energy crops e.g., silage maize (92%) and pig slurry (8%)	4035	[67]
7	Animal manure and energy crops	4000	[68]
8	Biowaste	2457–3950	[69]
9	Sewage sludge	2700–3800	[70]
10	Sewage sludge + Acid cheese whey	2800–3750	[70]
11	Dairy manure	3007	[71]
12	Biowaste and kitchen refuse	1010–2780	[72]
13	Municipal wastewater	2667	[73]

S/N	Feedstock	N Value [mg/L]	Reference
14	Maize silage and distillery stillage	2620	[74]
15	Poultry litter	1570–2473	[75]
16	Bio-slurry	170–2240	[76]
17	Source separated household waste	2200	[77]
18	Municipal solid waste	1308–1569	[78]
19	Swine manure	1135	[79]
20	Waste activated sludge and organic fraction of municipal solid waste	425–850	[80]
21	Sewage sludge (half-synthetic)	820	[81]
22	Piggery farm effluent	774	[82]
23	Yeast production wastewater	703	[83]
24	Cattle slurry and glycerin	600	[84]
25	Municipal wastewater sludge	280–590	[85]
26	**Cassava peeling residue (CPR)**	**573**	**[41]**
27	Sewage sludge and organic fraction of municipal solid waste	355–535	[86]
28	Municipal wastewater	435–520	[87]
29	Swine wastewater	460	[88]
30	Starch processing wastewater	240–383	[89]
31	Starch processing wastewater	265	[90]
32	Piggery wastewater	139	[91]

Table 2.
Comparison of nitrogen (N) content of liquid fraction of digestate derived from various feedstocks.

S/N	Feedstock	P Value [mg/L]	Reference
1	Pig slurry	800–1700	[65]
2	Dairy cow slurry	200–1000	[65]
3	Dairy manure	802	[63]
4	Sewage sludge	590–680	[70]
5	Sewage sludge + Acid cheese whey	500–550	[70]
6	Pig manure	492	[92]
7	Energy crops (92%) and pig slurry (8%)	412	[67]
8	Municipal wastewater	381	[73]
9	Bio-slurry	56–320	[76]
10	Cattle & pig slurries (main feedstocks), various food wastes (co-substrates)	292–315	[61]
11	Maize silage and distillery stillage	270	[74]
12	Fruit and vegetable food waste	261	[93]
13	Source separated household waste	230	[77]
14	Poultry litter	154–214	[75]
15	Piggery wastewater	185	[91]
16	Municipal wastewater sludge	100–185	[85]

S/N	Feedstock	P Value [mg/L]	Reference
17	Organic waste (Kitchen garbage, spoilt food, etc.)	58–167	[66]
18	Sewage sludge (half-synthetic)	130	[81]
19	Sewage sludge and organic fraction of municipal solid waste	29–120	[86]
20	Swine manure	115	[88]
21	Waste activated sludge and organic fraction of municipal solid waste	95	[80]
22	Municipal solid waste	60–62	[78]
23	Biowaste	35–55	[69]
24	Municipal wastewater	43	[94]
25	Starch processing wastewater	23–40	[89]
26	**Cassava peeling residue (CPR)**	**31**	**[41]**
27	Starch processing wastewater	28	[90]
28	Swine manure	25	[79]
29	Algal biomass (*Tetraselmis* sp.)	7	[95]
30	Yeast production wastewater	7	[83]

Table 3.
Comparison of phosphorus (P) content of liquid fraction of digestate derived from various feedstocks.

S/N	Feedstock	K Value [mg/L]	Reference
1	Animal manure and energy crops	3500	[68]
2	Pig manure	3258	[92]
3	Cattle & pig slurries (main feedstocks), various food wastes (co-substrates)	1337–2850	[64]
4	Organic fraction of municipal solid waste	700–2216	[96]
5	Poultry litter	1632–2100	[75]
6	Cattle slurry and 10% orange peel residue	1200	[84]
7	Source separated household waste	1130	[77]
8	Cattle slurry and 5% orange peel residue	1100	[84]
9	**Cassava peeling residue (CPR)**	**1066**	**[41]**
10	Baker's yeast industry wastewater	827	[97]
11	Swine manure	809	[79]
12	Cattle slurry and glycerin	800	[84]
13	Bio-slurry	100–434	[76]
14	Starch processing wastewater	102–176	[89]
15	Starch processing wastewater	174	[90]
16	Sewage sludge and organic fraction of municipal solid waste	28–33	[86]
17	Waste activated sludge and organic fraction of municipal solid waste	30	[80]

Table 4.
Comparison of potassium (K) content of liquid fraction of digestate derived from various feedstocks.

4. Estimation of fertilizer credit for liquid fraction (LF) of cassava peeling residue (CPR) digestate derived from one metric ton (1000 kg) cassava root

The nutrient values presented in **Table 1** are for digestate derived from 800 g CPR accumulated in the 3 L working volume of AD reactor [41, 61]. Therefore, total nutrient credits for the 800 g CPR are:

N = 573 mg/L × 3 L = 1719 mg (1.719 g).

P = 31 mg/L × 3 L = 93 mg (0.093 g).

K = 1066 mg/L × 3 L = 3198 mg (3.198 g).

With the nutrients credit for 800 g (0.8 kg) CPR established, estimation of corresponding nutrient credit for CPR generated from 1000 kg cassava root becomes possible. It has been reported that CPR constitutes about 19% mass fraction of fresh cassava root [98]. Consequently, 1000 kg cassava root would yield 190 kg CPR. Hence, N, P, and K fertilizer credits for CPR generated from 1000 kg cassava root are estimated as:

N = 190 kg/0.8 kg × 1.719 g.

P = 190 kg/0.8 kg × 0.093 g.

K = 190 kg/0.8 kg × 3.198 g.

The results are presented in **Table 5**

Nutrient	Quantity in LF of CPR digestate from 1000 kg cassava root	
	[g]	[kg]
Nitrogen (N)	408.2625	0.4082625
Phosphorus (P)	22.0875	0.0220875
Potassium (K)	759.525	0.759525

Table 5.
Nutrient credit for LF of digestate of CPR derived from 1000 kg cassava root.

5. Capability of liquid fraction (LF) of cassava peeling residue (CPR) digestate to supplant chemical fertilizer in cassava root production

Cassava crop is forbearing of harsh growing conditions such as drought, acidic soil, marginal land, varied elevation, swings of temperature and rainfall [99, 100]. However, research has shown that cassava is also responsive to adequate soil fertility and fertilizer application [101–103]. The equivalent root productivities in response to three cases of chemical fertilizer input are presented in **Table 6**.

Case	Fertilizer Input [kg/ha]			Root Output [kg/ha]
	Nitrogen (N)	Phosphorus (P_2O_5)	Potassium (K_2O)	
A	1.78	0.44	2.28	1000
B	1.81	1.03	3.30	1000
C	1.38	0.51	4.38	1000

Table 6.
Nutrient requirements for cassava root production (Derived from ref. [102]).

From the atomic weights of P, K and O, the elemental nutrient equivalent of the oxide forms (P_2O_5 and K_2O) could be computed with the equations:

$$P = 0.437\ (P2O5) \tag{1}$$

$$K = 0.830\ (K2O) \tag{2}$$

Consequently, the total P and total K corresponding to the total N required for the three cases of cassava root production outlined in **Table 6** are estimated and presented in **Table 7**.

Based on nutrients required to produce one metric ton (1000 kg) of cassava root shown in **Table 7**, and the nutrient credit for LF of digestate of CPR generated from 1000 kg cassava root presented in **Table 5**, the capability of LF of CPR digestate to supplant chemical fertilizer in cassava root production is estimated and outlined in **Table 8**. The proportion of production nutrient substituted range from about 23–30% for nitrogen; 5–11% for phosphorus; and 21–40% for potassium.

Case	Fertilizer Input [kg/ha]			Root Output [kg/ha]
	Nitrogen (N)	Phosphorus (P)	Potassium (K)	
A	1.78	0.19228	1.8924	1000
B	1.81	0.45011	2.739	1000
C	1.38	0.22287	3.6354	1000
Mean	1.6567	0.2884	2.7556	1000

Table 7.
Elemental nutrient requirements for cassava root production (Derived from ***Table 6****).*

6. Cost analysis

From the mean nutrient values in **Table 8**, about 25, 8, and 28% of N, P, and K respectively required for production of 1000 kg cassava root, and sourced from inorganic fertilizers are supplanted by liquid fraction of CPR digestate. The cost implications are analyzed and presented in **Table 9**. The analyses indicate that about 25% of the total financial cost of inorganic fertilizers is supplanted by liquid fraction of CPR digestate (**Table 9**).

Case	Nutrient	Nutrient required for production of 1000 kg cassava root (From: Table 7) [kg]	Nutrient credit for liquid fraction (LF) of digestate of CPR generated from 1000 kg cassava root (From: Table 5) [kg]	Proportion of nutrient required for production of 1000 kg cassava root supplanted by LF of digestate of CPR generated from 1000 kg cassava root [%]
A	Nitrogen (N)	1.78	0.408	22.92
	Phosphorus (P)	0.192	0.022	11.46
	Potassium (K)	1.892	0.760	40.17
B	Nitrogen (N)	1.81	0.408	22.54
	Phosphorus (P)	0.450	0.022	4.89
	Potassium (K)	2.739	0.760	27.75

Case	Nutrient	Nutrient required for production of 1000 kg cassava root (From: Table 7) [kg]	Nutrient credit for liquid fraction (LF) of digestate of CPR generated from 1000 kg cassava root (From: Table 5) [kg]	Proportion of nutrient required for production of 1000 kg cassava root supplanted by LF of digestate of CPR generated from 1000 kg cassava root [%]
C	Nitrogen (N)	1.38	0.408	29.56
	Phosphorus (P)	0.223	0.022	9.87
	Potassium (K)	3.635	0.760	20.91
Mean	Nitrogen (N)	1.6567	0.408	24.63
	Phosphorus (P)	0.2884	0.022	7.63
	Potassium (K)	2.7556	0.760	27.58

Table 8.
*Capability of liquid fraction of CPR digestate to supplant chemical fertilizers required for cassava root production (Estimate based on **Tables** 5 and 7).*

Variable	Unit	Nutrient			Total
		Nitrogen (N)	Phosphorus (P)	Potassium (K)	
Unit cost of nutrient*	US$/kg	1.34	3.95	2.33	—
Nutrient required for production of 1000 kg cassava root	kg	1.6567	0.2884	2.7556	—
Cost of nutrient required for production of 1000 kg cassava root	US$	2.22	1.1392	6.4205	9.7797
Nutrient from liquid fraction of digestate of CPR generated from 1000 kg cassava root	kg	0.408	0.022	0.760	—
Cost credit of nutrient from liquid fraction of digestate of CPR generated from 1000 kg cassava root	US$	0.5467	0.0869	1.7708	2.4044
Proportion of cost of nutrient required for production of 1000 kg cassava root saved by liquid fraction of CPR digestate	%	24.63	7.63	27.58	24.59

**Unit cost of liquid fertilizer derived from price data supplied by Ramsdell F&M Ltd. Brookings, SD USA. (Price as at 13th September 2021).*

Table 9.
Cost implications of supplanting chemical fertilizers with liquid fraction of CPR digestate in cassava root production.

7. Global fertilizer savings from liquid fraction (LF) of cassava peeling residue (CPR) digestate

In 2019, a total of 96 recorded countries/territories produced about 303.569 x 10^9 kg cassava root globally. The output ranged from 5000 kg for Maldives to 59.194 × 10^9 kg for Nigeria [104]. At 19% CPR mass fraction composition, 57.678 × 10^9 kg of CPR would be generated from the global root output. This quantity of CPR could be transformed to biogas and digestate via AD. Whole digestate could be separated into liquid and solid fractions using appropriate technologies [41]. The liquid fraction of CPR digestate could then be utilized to supplant inorganic fertilizers required for cassava root production. The cost data generated and presented in **Table 9** are applied to estimate the pecuniary value of global fertilizer savings from liquid fraction of CPR digestate substitution of chemical fertilizers. The results for each of the 96 countries/territories that produced cassava root in 2019 are presented in **Table 10**. Total global cost savings is about US$ 141.019 (€ 121.130) million. The range is from US$ 2.323 (€ 1.995) for Maldives, to US$ 27.498 (€ 23.620) million for Nigeria.

S/N	Country	2019 Cassava root output [x 10^9 kg][a]	CPR generated from 2019 root output @ 19% CPR mass fraction [x 10^9 kg]	Cost of chemical fertilizer required for 2019 root output [x 10^6 US$]	Potential savings from liquid fraction digestate derived from 2019 CPR output. (≈ 25% total fertilizer cost) [× 10^6 US$]
1	Nigeria	59.193708	11.24680452	109.9903742	27.49759354
2	Democratic Republic of the Congo	40.050112	7.60952128	74.41883526	18.60470882
3	Thailand	31.079966	5.90519354	57.75102126	14.43775532
4	Ghana	22.447635	4.26505065	41.71091584	10.42772896
5	Brazil	17.497115	3.32445185	32.51214176	8.128035439
6	Indonesia	14.586693	2.77147167	27.10416149	6.776040373
7	Cambodia	13.737921	2.61020499	25.52702174	6.381755435
8	Viet Nam	10.105224	1.91999256	18.77695124	4.69423781
9	Angola	9.000432	1.71008208	16.72408972	4.181022429
10	United Republic of Tanzania	8.184093	1.55497767	15.20721512	3.80180378
11	Cameroon	6.092549	1.15758431	11.32082728	2.830206819
12	Malawi	5.667887	1.07689853	10.53174455	2.632936138
13	Côte d'Ivoire	5.238244	0.99526636	9.733406421	2.433351605
14	China	4.986557	0.94744583	9.265735984	2.316433996
15	India	4.976	0.94544	9.246119568	2.311529892
16	China, mainland	4.975472	0.94533968	9.245138468	2.311284617
17	Sierra Leone	4.588612	0.87183628	8.526297268	2.131574317
18	Zambia	4.036584	0.76695096	7.500550304	1.875137576
19	Mozambique	3.987446	0.75761474	7.409244873	1.852311218
20	Benin	3.894777	0.74000763	7.237052619	1.809263155
21	Paraguay	3.384	0.64296	6.287955912	1.571988978
22	Madagascar	2.913862	0.55363378	5.414372278	1.35359307
23	Uganda	2.841625	0.53990875	5.280145602	1.320036401
24	Philippines	2.6308	0.499852	4.888402604	1.222100651
25	Burundi	2.408958	0.45770202	4.476188445	1.119047111
26	Lao People's Democratic Republic	2.258702	0.42915338	4.19699131	1.049247828
27	Guinea	2.145484	0.40764196	3.986616076	0.996654019

S/N	Country	2019 Cassava root output [x 10^9 kg][a]	CPR generated from 2019 root output @ 19% CPR mass fraction [x 10^9 kg]	Cost of chemical fertilizer required for 2019 root output [x 10^6 US$]	Potential savings from liquid fraction digestate derived from 2019 CPR output. (≈ 25% total fertilizer cost) [× 10^6 US$]
28	Congo	1.457028	0.27683532	2.707366379	0.676841595
29	Peru	1.286013	0.24434247	2.389596054	0.597399013
30	Rwanda	1.181825	0.22454675	2.195999851	0.548999963
31	Togo	1.11788	0.2123972	2.077180897	0.519295224
32	Senegal	1.030592	0.19581248	1.914987311	0.478746828
33	Colombia	1.026643	0.19506217	1.907649504	0.476912376
34	Kenya	0.970587	0.18441153	1.80348944	0.45087236
35	Cuba	0.795748	0.15119212	1.478613576	0.369653394
36	Central African Republic	0.730362	0.13876878	1.357117038	0.339279259
37	South Sudan	0.572531	0.10878089	1.06384447	0.265961117
38	Liberia	0.558222	0.10606218	1.037256302	0.259314075
39	Niger	0.513671	0.09759749	0.954474173	0.238618543
40	Haiti	0.507856	0.09649264	0.943669071	0.235917268
41	Venezuela (Bolivarian Republic of)	0.42162	0.0801078	0.783430252	0.195857563
42	Myanmar	0.392443	0.07456417	0.729215213	0.182303803
43	Gabon	0.337209	0.06406971	0.626582543	0.156645636
44	Chad	0.296976	0.05642544	0.551823876	0.137955969
45	Sri Lanka	0.281075	0.05340425	0.522277544	0.130569386
46	Zimbabwe	0.253835	0.04822865	0.471661728	0.117915432
47	Nicaragua	0.220786	0.04194934	0.41025196	0.10256299
48	Bolivia (Plurinational State of)	0.203327	0.03863213	0.377810642	0.09445266
49	Argentina	0.195852	0.03721188	0.363921023	0.090980256
50	Dominican Republic	0.17469	0.0331911	0.324599001	0.08114975
51	Costa Rica	0.159861	0.03037359	0.297044598	0.07426115
52	Papua New Guinea	0.155145	0.02947755	0.288281596	0.072070399
53	Somalia	0.093717	0.01780623	0.174139588	0.043534897
54	Equatorial Guinea	0.079646	0.01513274	0.147993657	0.036998414
55	Fiji	0.07603	0.0144457	0.141274612	0.035318653
56	Mali	0.070312	0.01335928	0.130649751	0.032662438
57	Ecuador	0.069863	0.01327397	0.129815444	0.032453861
58	Comoros	0.065071	0.01236349	0.120911223	0.030227806

S/N	Country	2019 Cassava root output [x 10^9 kg][a]	CPR generated from 2019 root output @ 19% CPR mass fraction [x 10^9 kg]	Cost of chemical fertilizer required for 2019 root output [x 10^6 US$]	Potential savings from liquid fraction digestate derived from 2019 CPR output. (≈ 25% total fertilizer cost) [× 10^6 US$]
59	Guinea-Bissau	0.056073	0.01065387	0.104191652	0.026047913
60	Malaysia	0.042267	0.00803073	0.07853813	0.019634533
61	El Salvador	0.029148	0.00553812	0.054161152	0.013540288
62	Mexico	0.027153	0.00515907	0.050454157	0.012613539
63	Honduras	0.026732	0.00507908	0.049671879	0.01241797
64	Jamaica	0.026529	0.00504051	0.049294676	0.012323669
65	Timor-Leste	0.021533	0.00409127	0.040011393	0.010002848
66	Guyana	0.01855	0.0035245	0.034468553	0.008617138
67	Panama	0.017234	0.00327446	0.032023236	0.008005809
68	Gambia	0.013174	0.00250306	0.024479176	0.006119794
69	China, Taiwan Province of	0.011085	0.00210615	0.020597515	0.005149379
70	Micronesia (Federated States of)	0.00842	0.0015998	0.015645564	0.003911391
71	Suriname	0.007783	0.00147877	0.014461927	0.003615482
72	Tonga	0.006692	0.00127148	0.012434693	0.003108673
73	Cabo Verde	0.005124	0.00097356	0.009521125	0.002380281
74	Guatemala	0.004185	0.00079515	0.007776328	0.001944082
75	Burkina Faso	0.004046	0.00076874	0.007518047	0.001879512
76	French Polynesia	0.003937	0.00074803	0.007315509	0.001828877
77	Brunei Darussalam	0.003382	0.00064258	0.00628424	0.00157106
78	Solomon Islands	0.003381	0.00064239	0.006282381	0.001570595
79	Trinidad and Tobago	0.002355	0.00044745	0.004375927	0.001093982
80	Saint Lucia	0.001459	0.00027721	0.002711031	0.000677758
81	Sao Tome and Principe	0.001384	0.00026296	0.00257167	0.000642917
82	Dominica	0.001277	0.00024263	0.002372849	0.000593212
83	New Caledonia	0.000832	0.00015808	0.001545975	0.000386494
84	Belize	0.000725	0.00013775	0.001347154	0.000336788
85	Cook Islands	0.000718	0.00013642	0.001334147	0.000333537
86	Mauritius	0.000715	0.00013585	0.001328572	0.000332143
87	Saint Vincent and the Grenadines	0.000586	0.00011134	0.001088872	0.000272218
88	Barbados	0.000486	0.00009234	0.000903057	0.000225764

S/N	Country	2019 Cassava root output [x 10^9 kg][a]	CPR generated from 2019 root output @ 19% CPR mass fraction [x 10^9 kg]	Cost of chemical fertilizer required for 2019 root output [x 10^6 US$]	Potential savings from liquid fraction digestate derived from 2019 CPR output. (≈ 25% total fertilizer cost) [× 10^6 US$]
89	Samoa	0.000474	0.00009006	0.00088076	0.00022019
90	Seychelles	0.000236	0.00004484	0.000438522	0.00010963
91	Grenada	0.000235	0.00004465	0.000436664	0.000109166
92	Bahamas	0.000203	0.00003857	0.000377203	9.43008E-05
93	Puerto Rico	0.00017	0.0000323	0.000315884	7.89711E-05
94	Antigua and Barbuda	0.000159	0.00003021	0.000295445	7.38612E-05
95	Niue	0.000044	0.00000836	8.17583E-05	2.04396E-05
96	Maldives	0.000005	0.00000095	9.29072E-06	2.32268E-06
97	**World Total**	**303.568814**	**57.67807466**	**564.0742668**	**141.0185667**

[a]*Data source: (Ref. [104]).*

Table 10.
Global fertilizer savings from liquid fraction of CPR digestate.

8. Conclusion

Haber-Bosch process facilitated the existence of inorganic fertilizers that revolutionized crop yield, improved nutrition, and enhanced food security. However, external costs associated with the production and application of inorganic fertilizers in agriculture are prodigious. Air quality degradation, climate perturbation, eutrophication, harmful algal blooms, ocean dead zones, pollution of surface water bodies and groundwater aquifers used as potable water supply sources are notable examples. Anaerobic digestion in the context of circular economic paradigm could provide viable solution. Anaerobic digestion transforms organic wastes and residues to beneficial biogas fuel and digestate biofertilizer. This chapter analyzed and presented the cost implications of liquid fraction of cassava peeling residue (CPR) digestate to supplant chemical fertilizers required for cassava root production. About 25% of fund used to purchase the chemical fertilizers required for cassava root production could be saved with the use of liquid fraction of CPR digestate. The global pecuniary saving is estimated at US$ 0.141 (€ 0.121) billion for year 2019 cassava root output. Thus, exploitation of liquid fraction of CPR digestate would save 25% pecuniary cost of inorganic fertilizers required for cassava root production, as well as attenuate afore mentioned external costs correlated with the production and application of the inorganic fertilizers.

Perspectives: There is severe scarcity of data on the speciation of nutrients content of digestates derived from anaerobic digestion of CPR as single feedstock. The few studies reported in available literature focused on biogas potentials of CPR co-digested with other substrates such as animal manures. The reports did not indicate any data on generated digestate. For future perspectives, experimental questions could address systematic studies to characterize the nutrient speciation in digestates derived from CPR as mono feedstock. Findings may not only corroborate the fertilizer values of CPR digestate reported in this chapter, but also establish CPR's nutrients influence and contribution when co-digested with other feedstocks. Furthermore, the work for this chapter searched, and could not find any study on the effects of CPR digestate on crop performance. Agronomic experiments designed with CPR digestate as bio-fertilizer, would provide valuable knowledge and insight on the suitability and practical impact of CPR digestate on yield and other performance indicators for cassava, and perhaps other crops.

Author details

Sammy N. Aso[1]*, Simeon C. Achinewhu[2] and Madu O. Iwe[3]

1 Food Engineering Laboratory, Rivers State University, Port Harcourt, Nigeria

2 Department of Food Science and Technology, Rivers State University, Port Harcourt, Nigeria

3 Department of Food Science and Technology, Michael Okpara University of Agriculture, Umudike, Nigeria

*Address all correspondence to: sammyasso@yahoo.com

References

[1] Haber F. The synthesis of ammonia from its elements. Stockholm, Sweden: Nobel Lecture. 1920 Available at: https://www.ias.ac.in/public/Volumes/reso/007/09/0086-0094.pdf

[2] Stewart WM, Dibb DW, Johnston AE, Smyth TJ. The contribution of commercial fertilizer nutrients to food production. Agronomy Journal. 2005;**97**(1):1-6. DOI: 10.2134/agronj2005.0001

[3] Erisman JW, Sutton MA, Galloway J, Klimont Z, Winiwarter W. How a century of ammonia synthesis changed the world. Nature Geoscience. 2008; **1**(10):636-639. DOI: 10.1038/ngeo325

[4] Vitousek PM, Aber JD, Howarth RW, Likens GE, Matson PA, Schindler DW, et al. Human alteration of the global nitrogen cycle: sources and consequences. Ecological Applications. 1997;**7**(3):737-750. DOI: 10.1890/1051-0761(1997)007[0737:HAOTGN]2.0.CO;2

[5] Dobermann A, Cassman KG. Cereal area and nitrogen use efficiency are drivers of future nitrogen fertilizer consumption. Science in China Series C: Life Sciences. 2005;**48**(2):745-758. DOI: 10.1007/BF03187115

[6] Dodds WK, Bouska WW, Eitzmann JL, Pilger TJ, Pitts KL, Riley AJ, et al. Eutrophication of U.S. freshwaters: Analysis of potential economic damages. Environmental Science and Technology. 2009;**43**(1):12-19. DOI: 10.1021/es801217q

[7] Exner ME, Hirsh AJ, Spalding RF. Nebraska's groundwater legacy: Nitrate contamination beneath irrigated cropland. Water Resources Research. 2014;**50**(5):4474-4489. DOI: 10.1002/2013WR015073

[8] Bommelaer O, Devaux J. Assessing Water Pollution Costs of Farming in France. Paris, France: Department of the commissioner general for sustainable development, No. 52 EV (English version); 2011. Available from: http://temis.documentation.developpement-durable.gouv.fr/docs/Temis/0070/Temis-0070550/19342_ENG.pdf

[9] GEF: Secretariat of the Convention on Biological Diversity and the Scientific and Technical Advisory Panel. Impacts of marine debris on biodiversity: Current status and potential solutions. CBD Technical Series No. 67. Montreal, Quebec, Canada: Secretariat of the Convention on Biological Diversity; 2012. 61pp. Available from: https://www.cbd.int/doc/publications/cbd-ts-67-en.pdf

[10] Portmann RW, Daniel JS, Ravishankara AR. Stratospheric ozone depletion due to nitrous oxide: Influences of other gases. Philosophical Transactions of the Royal Society B: Biological Sciences. 2012;**367**(1593):1256-1264. DOI: 10.1098/rstb.2011.0377

[11] Bouwman A, Boumans LJM. Global estimates of gaseous emissions of NH_3, NO and N_2O from agricultural land. Rome: International Fertilizer Industry Association (IFA) and the Food and Agriculture Organization of the United Nations (FAO); 2001

[12] Artíñano B, Pujadas M, Alonso-Blanco E, Becerril-Valle M, Coz E, Gómez-Moreno FJ, et al. Real-time monitoring of atmospheric ammonia during a pollution episode in Madrid (Spain). Atmospheric Environment. 2018;**189**:80-88. DOI: 10.1016/j.atmosenv.2018.06.037

[13] Van Damme M, Clarisse L, Whitburn S, Hadji-Lazaro J, Hurtmans D, Clerbaux C, et al. Industrial and agricultural ammonia point sources exposed. Nature. 2018; **564**(7734):99-103. DOI: 10.1038/s41586-018-0747-1

[14] Shephard MW, Dammers E, Cady-Pereira KE, Kharol SK, Thompson J, Gainariu-Matz Y, et al. Ammonia measurements from space with the Cross-track Infrared Sounder: Characteristics and applications. Atmospheric Chemistry and Physics. 2020;**20**(4):2277-2302. DOI: 10.5194/acp-20-2277-2020

[15] Verdi L, Kuikman PJ, Orlandini S, Mancini M, Napoli M, Dalla Marta A. Does the use of digestate to replace mineral fertilizers have less emissions of N_2O and NH_3? Agricultural and Forest Meteorology. 2019;**269**:112-118. DOI: 10.1016/j.agrformet.2019.02.004

[16] EEA: European Environment Agency. Air quality in Europe — 2015 report. Copenhagen, Denmark: European Environment Agency; 2015

[17] EEA: European Environment Agency. Air quality in Europe — 2020 report. Copenhagen, Denmark: European Environment Agency; 2020

[18] Good AG, Beatty PH. Fertilizing nature: A tragedy of excess in the commons. PLoS Biology. 2011;**9**(8): e1001124. DOI: 10.1371/journal.pbio.1001124

[19] Mitsch WJ, Day JW, Gilliam JW, Groffman PM, Hey DL, Randall GW, et al. Reducing nitrogen loading to the gulf of Mexico from the Mississippi river basin: Strategies to counter a persistent ecological problem. Bioscience. 2001;**51**(5):373-388. DOI: 10.1641/0006-3568(2001)051[0373:RNLTTG]2.0.CO;2

[20] Young CE, Magleby RS. Agricultural pollution control: Implications from the rural clean water program 1. JAWRA Journal of the American Water Resources Association. 1987;**23**(4):701-707. DOI: 10.1111/j.1752-1688.1987.tb00844.x

[21] Boesch DF, Brinsfield RB, Magnien RE. Chesapeake Bay eutrophication: Scientific understanding, ecosystem restoration, and challenges for agriculture. Journal of Environmental Quality. 2001;**30**(2): 303-320. DOI: 10.2134/jeq2001.302303x

[22] Ju XT, Xing GX, Chen XP, Zhang SL, Zhang LJ, Liu XJ, et al. Reducing environmental risk by improving N management in intensive Chinese agricultural systems. Proceedings of the National Academy of Sciences. 2009;**106**(9):3041-3046. DOI: 10.1073/pnas.0813417106

[23] Zhai Y, Zhao X, Teng Y, Li X, Zhang J, Wu J, et al. Groundwater nitrate pollution and human health risk assessment by using HHRA model in an agricultural area, NE China. Ecotoxicology and Environmental Safety. 2017;**137**:130-142. DOI: 10.1016/j.ecoenv.2016.11.010

[24] Cavigelli MA, Del Grosso SJ, Liebig MA, Snyder CS, Fixen PE, Venterea RT, et al. US agricultural nitrous oxide emissions: Context, status, and trends. Frontiers in Ecology and the Environment. 2012;**10**(10):537-546. DOI: 10.1890/120054

[25] Decock C. Mitigating nitrous oxide emissions from corn cropping systems in the midwestern US: Potential and data gaps. Environmental Science & Technology. 2014;**48**(8):4247-4256. DOI: 10.1021/es4055324

[26] Lu Y, Huang Y, Zou J, Zheng X. An inventory of N_2O emissions from agriculture in China using precipitation-rectified emission factor and background emission. Chemosphere. 2006;**65**(11):1915-1924. DOI: 10.1016/j.chemosphere.2006.07.035

[27] USEPA: United States Environmental Protection Agency. An Urgent Call to Action – Report of the State-EPA Nutrient Innovations Task Group. Washington, DC: US EPA; 2009

[28] Velthof GL, Oudendag D, Witzke HP, Asman WAH, Klimont Z, Oenema O. Integrated assessment of nitrogen losses from agriculture in EU-27 using MITERRA-EUROPE. Journal of Environmental Quality. 2009;**38**(2): 402-417. DOI: 10.2134/jeq2008.0108

[29] MacDonald GK, Bennett EM, Potter PA, Ramankutty N. Agronomic phosphorus imbalances across the world's croplands. Proceedings of the National Academy of Sciences. 2011; **108**(7):3086-3091. DOI: 10.1073/pnas.1010808108

[30] Bommelaer O, Devaux J. Financing Water Resources Management in France January 2012 update. Paris, France: Department of the commissioner general for sustainable development; 2012

[31] Sutton MA, Oenema O, Erisman JW, Leip A, van Grinsven H, Winiwarter W. Too much of a good thing. Nature. 2011;**472**:159-161. DOI: 10.1038/472159a

[32] FAO: Food and Agriculture Organization of the United Nations. FAOSTAT. Fertilizers by Nutrient. Rome: FAO; 2021

[33] Filka, P. Design of wasteless food processing plants. In: Food Engineering and Process Applications Vol. 2, Unit operations. Le Maguer, M., & Jelen, P. Eds. Elsevier Applied Science Publishers, London and New York. 1986. Pp. 539-552

[34] Pearce D, Turner RK, O'riordan T, Adger N, Atkinson G, Brisson I, et al. Blueprint 3: Measuring sustainable development. London, UK: Earthscan Publications; 1993. 224pp

[35] Aso SN. Food engineering stratagem to protect the environment and improve the income opportunities of gari processors. Journal of Nigerian Environmental Society (JNES). 2004; **2**(1):31-36

[36] Ellen MacArthur Foundation. Towards the Circular Economy: Economic and Business Rationale for an Accelerated Transition. Hoboken, USA: Ellen MacArthur; 2013. Available from: https://www.ellenmacarthurfoundation.org/assets/downloads/publications/Ellen-MacArthur-Foundation-Towards-the-Circular-Economy-vol.1.pdf

[37] Jurgilevich A, Birge T, Kentala-Lehtonen J, Korhonen-Kurki K, Pietikäinen J, Saikku L, et al. Transition towards circular economy in the food system. Sustainability. 2016;**8**(1):69. DOI: 10.3390/su8010069

[38] Awasthi MK, Sarsaiya S, Patel A, Juneja A, Singh RP, Yan B, et al. Refining biomass residues for sustainable energy and bio-products: An assessment of technology, its importance, and strategic applications in circular bio-economy. Renewable and Sustainable Energy Reviews. 2020;**127**: 109876. DOI: 10.1016/j.rser.2020.109876

[39] Garlapati VK, Chandel AK, Kumar SJ, Sharma S, Sevda S, Ingle AP, et al. Circular economy aspects of lignin: Towards a lignocellulose biorefinery. Renewable and Sustainable Energy Reviews. 2020;**130**:109977. DOI: 10.1016/j.rser.2020.109977

[40] Awasthi AK, Cheela VS, D'Adamo I, Iacovidou E, Islam MR, Johnson M, et al. Zero waste approach towards a sustainable waste management. Resources, Environment and Sustainability. 2021;**3**:100014. DOI: 10.1016/j.resenv.2021.100014

[41] Aso SN. Digestate: The Coproduct of Biofuel Production in a Circular Economy, and New Results for Cassava Peeling Residue Digestate. [Online First]. London, UK: IntechOpen; 2020. DOI: 10.5772/intechopen.91340

[42] Alburquerque JA, Fuente C, Campoy M, Carrasco L, Nájera I,

Baixauli C, et al. Agricultural use of digestate for horticultural crop production and improvement of soil properties. European Journal of Agronomy. 2012;**43**:119-128. DOI: 10.1016/j.eja.2012.06.001

[43] Garcia-Sánchez M, Garcia-Romera I, Cajthaml T, Tlustoš P, Száková J. Changes in soil microbial community functionality and structure in a metal-polluted site: The effect of digestate and fly ash applications. Journal of Environmental Management. 2015;**162**: 63-73. DOI: 10.1016/j.jenvman.2015.07.042

[44] Walsh JJ, Jones DL, Edwards-Jones G, Williams AP. Replacing inorganic fertilizer with anaerobic digestate may maintain agricultural productivity at less environmental cost. Journal of Plant Nutrition and Soil Science. 2012;**175**(6): 840-845. DOI: 10.1002/jpln.201200214

[45] Sigurnjak I, Vaneeckhaute C, Michels E, Ryckaert B, Ghekiere G, Tack FMG, et al. Fertilizer performance of liquid fraction of digestate as synthetic nitrogen substitute in silage maize cultivation for three consecutive years. Science of the Total Environment. 2017;**599–600**:1885-1894. DOI: 10.1016/j.scitotenv.2017.05.120

[46] Duan GL, Zhang HM, Liu YX, Jia Y, Hu Y, Cheng WD. Long-term fertilization with pig-biogas residues results in heavy metal accumulation in paddy field and rice grains in Jiaxing of China. Soil Science and Plant Nutrition. 2012;**58**(5):637-646. DOI: 10.1080/00380768.2012.726597

[47] Kouřimská L, Poustková I, Babička L. The use of digestate as a replacement of mineral fertilizers for vegetables growing. Scientia Agriculturae Bohemica. 2012;**43**(4): 121-126. DOI: 10.7160/sab.2012.430401

[48] Tambone F, Orzi V, Zilio M, Adani F. Measuring the organic amendment properties of the liquid fraction of digestate. Waste Management. 2019;**88**:21-27. DOI: 10.1016/j.wasman.2019.03.024

[49] Coelho JJ, Hennessy A, Casey I, Woodcock T, Kennedy N. Biofertilisation with anaerobic digestates: Effects on the productive traits of ryegrass and soil nutrients. Journal of Soil Science and Plant Nutrition. 2020;**20**:1-14. DOI: 10.1007/s42729-020-00237-7

[50] Timonen K, Sinkko T, Luostarinen S, Tampio E, Joensuu K. LCA of anaerobic digestion: Emission allocation for energy and digestate. Journal of Cleaner Production. 2019;**235**: 1567-1579. DOI: 10.1016/j.jclepro.2019.06.085

[51] Alonso AM, Guzmán GJ. Comparison of the efficiency and use of energy in organic and conventional farming in Spanish agricultural systems. Journal of Sustainable Agriculture. 2010;**34**(3):312-338. DOI: 10.1080/10440041003613362

[52] Aguilera E, Guzmán Casado G, Infante Amate J, Soto Fernández D, García Ruiz R, Herrera A, et al. Embodied Energy in Agricultural Inputs. Incorporating a Historical Perspective. Girona, Spain: Spanish Society of Agrarian History. 2015 DT-SEHA 1507; http://hdl.handle.net/10234/141278

[53] Cerilles AWE. Gahung-Gahung organic cassava farming system: A climate change adaptive and poverty-alleviating farming strategy. Journal of Agricultural Technology. 2015;**11**(8): 1669-1675

[54] Favoino E, Hogg D. Effects of composted organic waste on ecosystems – A specific angle: The potential contribution of biowaste to tackle Climate Change and references to the soil policy. In: Fuchs JG, Kupper T, Tamm L, Schenk K, editors. Compost

and digestate: Sustainability, benefits, impacts for the environment and for plant production. Solothurn, Switzerland: Proceedings of the international congress CODIS 2008, February 27–29, 2008; 2008. pp. 145-156, Research Institute of Organic Agriculture FiBL, Ackerstrasse, CH-5070 Frick Available from: https://orgprints.org/13135/1/fuchs-etal-proceedings-codis-2008.pdf#page=151

[55] Aguilera E, Lassaletta L, Sanz-Cobena A, Garnier J, Vallejo A. The potential of organic fertilizers and water management to reduce N_2O emissions in Mediterranean climate cropping systems. A review. Agriculture, Ecosystems & Environment. 2013;**164**: 32-52. DOI: 10.1016/j.agee.2012.09.006

[56] Aguilera E, Lassaletta L, Gattinger A, Gimeno BS. Managing soil carbon for climate change mitigation and adaptation in Mediterranean cropping systems: A meta-analysis. Agriculture, Ecosystems & Environment. 2013;**168**:25-36. DOI: 10.1016/j.agee.2013.02.003

[57] Pimentel D, Harvey C, Resosudarmo P, Sinclair K, Kurz D, McNair M, et al. Environmental and economic costs of soil erosion and conservation benefits. Science. 1995; **267**(5201):1117-1123. DOI: 10.1126/science.267.5201.1117

[58] Bolduan R, Deller B, Kluge R, Mokry M, Flaig H. Influence of mid-term application of composts on chemical, physical and biological soil properties of agricultural soils in field trials of practical importance. In: Fuchs JG, Kupper T, Tamm L, Schenk K, editors. Compost and digestate: Sustainability, benefits, impacts for the environment and for plant production. Solothurn, Switzerland: Proceedings of the international congress CODIS 2008, February 27–29, 2008; 2008. pp. 91-92 Research Institute of Organic Agriculture FiBL, Ackerstrasse, CH-5070 Frick. Available from: https://orgprints.org/id/eprint/13135/contents#page=97

[59] Armengot L, José-María L, Chamorro L, Sans FX. Weed harrowing in organically grown cereal crops avoids yield losses without reducing weed diversity. Agronomy for Sustainable Development. 2013;**33**(2):405-411. DOI: 10.1007/s13593-012-0107-8

[60] Möller K, Müller T. Effects of anaerobic digestion on digestate nutrient availability and crop growth: A review. Engineering in Life Sciences. 2012;**12**(3):242-257. DOI: 10.1002/elsc.201100085

[61] Aso, S. N. Synergistic enzymatic hydrolysis of cassava starch and anaerobic digestion of cassava waste (Dissertation). University of Florida, Department of Agricultural and Biological Engineering, Gainesville, FL. USA; 2013. Available from: https://search.proquest.com/docview/1547360007/fulltextPDF/29A82F37217B4D11PQ/1?accountid=28594 OR @ https://search.proquest.com/docview/1547360007?pq-origsite=gscholar

[62] Svehla P, Caceres LMV, Michal P, Tlustos P. Nitrification of the liquid phase of digestate can help with the reduction of nitrogen losses. Environmental Technology & Innovation. 2020;**17**: 100514. DOI: 10.1016/j.eti.2019.100514

[63] Martin JH. A Comparison of dairy cattle manure management with and without anaerobic digestion and biogas utilization. Report for the AgSTAR program, U.S. environmental protection agency. 2003. 58 pp. Available from: https://www.ncgreenpower.org/documents/nydairy2003.pdf

[64] Kocatürk-Schumacher NP, Zwart K, Bruun S, Brussaard L, Jensen LS. Does the combination of biochar and

clinoptilolite enhance nutrient recovery from the liquid fraction of biogas digestate? Environmental Technology. 2017;**38**(10):1313-1323. DOI: 10.1080/09593330.2016.1226959

[65] Møller HB, Lund I, Sommer SG. Solid–liquid separation of livestock slurry: Efficiency and cost. Bioresource Technology. 2000;**74**(3):223-229. DOI: 10.1016/S0960-8524(00)00016-X

[66] Waeger F, Delhaye T, Fuchs W. The use of ceramic microfiltration and ultrafiltration membranes for particle removal from anaerobic digester effluents. Separation and Purification Technology. 2010;**73**(2):271-278. DOI: 10.1016/j.seppur.2010.04.013

[67] Piccoli I, Virga G, Maucieri C, Borin M. Digestate liquid fraction treatment with filters filled with recovery materials. Water. 2021;**13**(1): 21. DOI: 10.3390/w13010021

[68] Bauer A, Mayr H, Hopfner-Sixt K, Amon T. Detailed monitoring of two biogas plants and mechanical solid-liquid separation of fermentation residues. Journal of Biotechnology. 2009;**142**(1):56-63. DOI: 10.1016/j.jbiotec.2009.01.016

[69] Mayer M, Smeets W, Braun R, Fuchs W. Enhanced ammonium removal from liquid anaerobic digestion residuals in an advanced sequencing batch reactor system. Water Science and Technology. 2009;**60**(7):1649-1660. DOI: 10.2166/wst.2009.546

[70] Bis M. Effect of acid whey addition for sewage sludge co-digestion on the nitrogen and phosphorus release. Journal of Ecological Engineering. 2020;**21**(6):13-21. DOI: 10.12911/22998993/123161

[71] Levine RB, Costanza-Robinson MS, Spatafora GA. *Neochloris oleoabundans* grown on anaerobically digested dairy manure for concomitant nutrient removal and biodiesel feedstock production. Biomass and Bioenergy. 2011;**35**(1):40-49. DOI: 10.1016/j.biombioe.2010.08.035

[72] Graja S, Wilderer PA. Characterization and treatment of the liquid effluents from the anaerobic digestion of biogenic solid waste. Water Science and Technology. 2001;**43**(3): 265-274. DOI: 10.2166/wst.2001.0146

[73] Cai T, Park SY, Racharaks R, Li Y. Cultivation of *Nannochloropsis salina* using anaerobic digestion effluent as a nutrient source for biofuel production. Applied Energy. 2013;**108**:486-492. DOI: 10.1016/j.apenergy.2013.03.056

[74] Kisielewska M, Dębowski M, Zieliński M, Kazimierowicz J, Quattrocelli P, Bordiean A. Effects of liquid digestate treatment on sustainable microalgae biomass production. Bioenergy Research. 2021:1-14. DOI: 10.1007/s12155-021-10251-x

[75] Singh M, Reynolds DL, Das KC. Microalgal system for treatment of effluent from poultry litter anaerobic digestion. Bioresource Technology. 2011;**102**(23):10841-10848. DOI: 10.1016/j.biortech.2011.09.037

[76] Pirelli T, Rossi A, Miller C. Sustainability of biogas and cassava-based ethanol value chains in Viet Nam: Results and recommendations from the implementation of the Global Bioenergy Partnership indicators, FAO Environment and Natural Resources Management Working Paper 69. Rome: FAO; 2018 Available from: http://www.fao.org/3/i9181en/I9181EN.pdf

[77] Haraldsen TK, Andersen U, Krogstad T, Sørheim R. Liquid digestate from anaerobic treatment of source-separated household waste as fertiliser to barley. Waste Management & Research. 2017;**29**:1271-1276. DOI: 10.1177/0734242X11411975

[78] Fricke K, Santen H, Wallmann R. Comparison of selected aerobic and

anaerobic procedures for MSW treatment. Waste Management. 2005; **25**(8):799-810. DOI: 10.1016/j.wasman.2004.12.018

[79] Cheng J, Xu J, Huang Y, Li Y, Zhou J, Cen K. Growth optimisation of microalga mutant at high CO_2 concentration to purify undiluted anaerobic digestion effluent of swine manure. Bioresource Technology. 2015; **177**:240-246. DOI: 10.1016/j.biortech.2014.11.099

[80] Frison N, Di Fabio S, Cavinato C, Pavan P, Fatone F. Best available carbon sources to enhance the via-nitrite biological nutrients removal from supernatants of anaerobic co-digestion. Chemical Engineering Journal. 2013;**215**: 15-22. DOI: 10.1016/j.cej.2012.10.094

[81] Ek M, Bergstrom R, Bjurhem J-E, Bjorlenius B, Hellstrom D. Concentration of nutrients from urine and reject water from anaerobically digested sludge. Water Science & Technology. 2006;**54**:437-444. DOI: 10.2166/wst.2006.924

[82] Ji F, Zhou Y, Pang A, Ning L, Rodgers K, Liu Y, et al. Fed-batch cultivation of *Desmodesmus* sp in anaerobic digestion wastewater for improved nutrient removal and biodiesel production. Bioresource Technology. 2015;**184**:116-122. DOI: 10.1016/j.biortech.2014.09.144

[83] Posadas E, Bochon S, Coca M, García-González MC, García-Encina PA, Muñoz R. Microalgae-based agro-industrial wastewater treatment: A preliminary screening of biodegradability. Journal of Applied Phycology. 2014;**26**:2335-2345. DOI: 10.1007/s10811-014-0263-0

[84] Alburquerque JA, Fuente C, Ferrer-Costa A, Carrasco L, Cegarra J, Abad M, et al. Assessment of the fertiliser potential of digestates from farm and agroindustrial residues. Biomass and Bioenergy. 2012;**40**:181-189. DOI: 10.1016/j.biombioe.2012.02.018

[85] Hollinshead WD, Varman AM, You L, Hembree Z, Tang YJ. Boosting D-lactate production in engineered cyanobacteria using sterilized anaerobic digestion effluents. Bioresource Technology. 2014;**169**:462-467. DOI: 10.1016/j.biortech.2014.07.003

[86] Fatone F, Dante M, Nota E, Di Fabio S, Frison N, Pavan P, et al. Biological short-cut nitrogen removal from anaerobic digestate in a demonstration sequencing batch reactor. Chemical Engineering Transactions. 2011;**24**:1135-1140

[87] Ahn YH, Choi HC. Autotrophic nitrogen removal from sludge digester liquids in upflow sludge bed reactor with external aeration. Process Biochemistry. 2006;**41**(9):1945-1950. DOI: 10.1016/j.procbio.2006.04.006

[88] Cao L, Wang J, Xiang S, Huang Z, Ruan R, Liu Y. Nutrient removal from digested swine wastewater by combining ammonia stripping with struvite precipitation. Environmental Science and Pollution Research. 2019; **26**(7):6725-6734. DOI: 10.1007/s11356-019-04153-x

[89] Tan X, Chu H, Zhang Y, Yang L, Zhao F, Zhou X. *Chlorella pyrenoidosa* cultivation using anaerobic digested starch processing wastewater in an airlift circulation photobioreactor. Bioresource Technology. 2014;**170**: 538-548. DOI: 10.1016/j.biortech.2014.07.086

[90] Yang L, Tan X, Li D, Chu H, Zhou X, Zhang Y, et al. Nutrients removal and lipids production by *Chlorella pyrenoidosa* cultivation using anaerobic digested starch wastewater and alcohol wastewater. Bioresource Technology. 2015;**181**: 54-61. DOI: 10.1016/j.biortech.2015.01.043

[91] Xu J, Zhao Y, Zhao G, Zhang H. Nutrient removal and biogas upgrading by integrating freshwater algae cultivation with piggery anaerobic digestate liquid treatment. Applied Microbiology and Biotechnology. 2015; **99**(15):6493-6501. DOI: 10.1007/s00253-015-6537-x

[92] Shi L, Xie S, Hu Z, Wu G, Morrison L, Croot P, et al. Nutrient recovery from pig manure digestate using electrodialysis reversal: Membrane fouling and feasibility of long-term operation. Journal of Membrane Science. 2019;**573**:560-569. DOI: 10.1016/j.memsci.2018.12.037

[93] Tuszynska A, Wilinska A, Czerwionka K. Phosphorus and nitrogen forms in liquid fraction of digestates from agricultural biogas plants. Environmental Technology. 2020;**42**:1-13. DOI: 10.1080/09593330.2020.1770339

[94] Wang H, Xiao K, Yang J, Yu Z, Yu W, Xu Q, et al. Phosphorus recovery from the liquid phase of anaerobic digestate using biochar derived from iron– rich sludge: A potential phosphorus fertilizer. Water Research. 2020;**174**:115629. DOI: 10.1016/j.watres.2020.115629

[95] Erkelens M, Ward AJ, Ball AS, Lewis DM. Microalgae digestate effluent as a growth medium for *Tetraselmis* sp. in the production of biofuels. Bioresource Technology. 2014;**167**:81-86. DOI: 10.1016/j.biortech.2014.05.126

[96] Macé S, Dosta J, Galí A, Mata-Alvarez J. Optimization of biological nitrogen removal via nitrite in a SBR treating supernatant from the anaerobic digestion of municipal solid wastes. Industrial & Engineering Chemistry Research. 2006;**45**(8):2787-2792. DOI: 10.1021/ie0509140

[97] Uysal A, Demir S, Sayilgan E, Eraslan F, Kucukyumuk Z. Optimization of struvite fertilizer formation from baker's yeast wastewater: Growth and nutrition of maize and tomato plants. Environmental Science and Pollution Research. 2014;**21**(5):3264-3274. DOI: 10.1007/s11356-013-2285-6

[98] Aso SN, Teixeira AA, Achinewhu SC. Cassava residues could provide sustainable bioenergy for cassava producing nations. Chapter 13. In: Waisundara VY, editor. Cassava. Rijeka, Croatia: IntechOpen; 2018. pp. 219-240. DOI: 10.5772/intechopen.72166

[99] Rogers DJ. Some botanical and ethnological considerations of *Manihot esculenta*. Economic Botany. 1965;**19**(4): 369-377

[100] CGIAR: Consultative Group on International Agricultural Research. Report on the Inter-Centre Review of Root and Tuber Crops Research in the CGIAR. Rome: FAO; 1997

[101] Wilson H, Ovid A. Influence of fertilizers on cassava production under rainfed conditions. Journal of Plant Nutrition. 1994;**17**(7):1127-1135. DOI: 10.1080/01904169409364793

[102] Imas P, John KS. Potassium Nutrition of Cassava - Research Findings. Zug, Switzerland: International Potash Institute (IPI); 2013 e-ifc No. 34

[103] Byju G, Suja G. Mineral nutrition of cassava. Advances in Agronomy. 2020;**159**:169-235. DOI: 10.1016/bs.agron.2019.08.005

[104] FAO: Food and Agriculture Organization of the United Nations. FAOSTAT. Crops and Livestock Products. Rome: FAO; 2021

Section 2

Case Studies and Evaluation

Chapter 6

A Case Study for Economic Viability of Biogas Production from Municipal Solid Waste in the South of Chile

Jean Pierre Doussoulin and Cristina Salazar Molina

Abstract

This research evaluated the technical and economic feasibility of a biogas plant in the south of Chile to generate energy (WtE) for the plant's own consumption, energy for sale to the country's electricity grid and produce biofertilizer from municipal solid waste (MSW). In the town of Panguipulli, 26 tons of solid waste are produced daily, of which 12 tons correspond to household organic waste. These arrive directly to a landfill, wasting their potential to generate products and energy. To study the economic feasibility, an analysis was carried out on the investment, costs and income that make up the cash flow of the project evaluated at 15 years. The results gave an NPV of 214.099.637 CLP and an IRR of 15% at a real discount rate of 10%, with a payback period of 6 years. The research concluded that it is feasible to design a biogas plant that works from household organic waste in Panguipulli. This will contribute to the mitigation of climate change and will promote circular economy actions and the sustainable management of MSW in the south of Chile.

Keywords: economic viability, biogas, municipal solid waste, Chile, waste to energy

1. Introduction

This chapter points out the problems present in cities as a result of the excessive growth of waste production. This negative externality must be managed through a management system, which not only accumulates it in a sanitary landfill, but also generates social and economic benefits. From this angle, Doussoulin highlights the important role of the state in supporting the transition from a linear production system to a circular one, urban waste is reused, extending its useful life and reducing negative externalities on the biosphere [1]. One of the key sectors in the generation of urban waste and its recycling is the construction sector in Latin American cities [2, 3].

Currently, there are several options to reuse urban waste. For example, composting, recycling and biomass that can be transformed into biogas, the latter topic will be addressed in the next investigation [2]. It will be necessary to understand some concepts about biogas. Rivas defines biogas production from household organic waste as a natural process, without oxygen carried out by micro-organisms, this involves the fermentation of organic materials to obtain the biogas [3]. Furthermore, biodigesters are systems designed to optimize the production of biogas, obtaining clean and low-cost energy [4]. As some authors have stated; gas extraction from waste, responds to the need to close the circle, returning natural resources to their origin [5]. The technological, legal and economic challenges and the opportunities for improvement in the well-being of developing countries have been studied and pointed out by various authors [6, 7]. The demonstration issues in major countries can be illustrated as follows (see **Table 1**).

Table 1 shows that the extraction of biogas from garbage is a relevant issue in South American countries. This is also emphasized by various Chilean authors on issues such as: the design of networks of biogas [26], environmental sustainability [27] and municipal waste management [28]. Therefore, this chapter continues and deepens these works taking advantage of the challenges and opportunities of biogas production. Thus, this study aims to study the feasibility of profitably investing in a biogas generating plant in the commune of Panguipulli from household organic waste. This will mean a crucial advance towards the reduction of the waste that reaches the landfill, therefore less environmental pollution, promotion of unconventional energies and direct solutions to citizens' problems by having a low-cost, good-quality product available. This research is mainly related to the search to alleviate energy poverty that currently exists, reducing economic barriers and in this way making a product as essential as gas more accessible to the public, whether it is used directly as fuel or electricity is generated from it [29]. This is why it is intended in the following research, to discover if it is feasible to invest in a generating plant of biogas in the Panguipulli commune by calculating the costs of the installation of a large-scale plant that meets the needs of the commune, as well as a calculation of the costs of the materials involved in the entire generation process of biogas, and finally to discover if the investment is recovered and if so, in how long a time.

The importance of this study concerns: first, the results will provide an important economic and time saving, since they will be of great help in upcoming projects related to landfill waste management policies and the generation of renewable energies in Chile. An attractive investment project in the medium and long term for

Country	Scholar	Issues
Belgium	[8, 9]	Anaerobic reactor for the biogas production from the pineapple and sweet sorghum
France	[10, 11]	Anaerobic reactor for the biogas production from the feedstock
UK	[12, 13]	Transport and energy crops fodder beet, forage maize, sugar beet and ryegrass
Argentina	[14, 15]	Biogas potential from MSW and aquatic plants
Brazil	[16, 17]	Biogas potential from MSW
USA	[18, 19]	Chemicals industry
South Africa	[20, 21]	Agricultural crops biogas
Canada	[22, 23]	Circular economy
China	[24, 25]	Household biogas use in the rural area

Data Source: [10–27].

Table 1.
Main demonstration issues related to biogas.

the entity that has the financial resources to carry it out. Second, this research will also carry out a study of the composition and volume of a substrate to be used in this specific case, investment analysis, costs and income that will make up a cash flow of a project evaluated at 15 years. In addition, some economic indicators are calculated to evaluate the viability of the project, these are: net present value (NPV), Internal rate of return (IRR) and Payback. Third, there is not much research on biogas plants in southern Chile. These biogas plants operate on a very small scale, the result of which is that there is no literature related to this geographical area.

This study explores the gap in the literature by answering whether the construction of a biogas plant in the commune of Panguipulli is economically profitable? The added value of this proposal is that it proposes an alternative use of the biogas applicable to the national reality and specifically to the Panguipulli commune, reducing negative environmental externalities as a result of their mismanagement emissions. Indeed, there is a lack of knowledge of the energetic potentiality of the biogas, for which an energetic waste arises and economical from the biogas emanating from the landfill. All of the above allow biogas generated in the sanitary landfill to not be managed correctly, causing the release of greenhouse gases such as CH_4 and CO_2 to a greater extent, which contribute to global warming, in addition to the contamination of the land and underground water.

Next, a compilation of information related to the biogas generation, similar studies, history of waste management and other data that the author considered relevant, all this was consulted in materials of authors with track records.

The chapter is structured as follows: Section 2 outlines a background of biogas production. Section 3 identifies the main results. Section 4 concludes and proposes future research direction.

2. Background

A large amount of waste is generated uncontrollably every day. From an environmental perspective, it is good to reduce the amount of waste that ends up to landfills, a part of this garbage being household organic waste that is usually thrown away along with everything else. In some parts of the world, the great potential that these projects have has been understood and projects have been created to reduce pollution, promote non-conventional renewable energies and generate a good quality product that allows an economic profit to be obtained. In other words, Parra refers to the fact that food residues (RA) have a high potential for reuse through biological processes such as anaerobic digestion (AD), especially due to their high content of biodegradable organic matter [30].

As a result of the decomposition of this organic material carried out by microorganisms, biogas is produced. A study by Gamma engineers defines it as a combustible gas that is generated in natural environments or specific devices, by the biodegradation reactions of organic matter, through the action of microorganisms in the absence of oxygen, that is, under anaerobic conditions [31]. Therefore, to optimize biogas production, this process is carried out in biodigesters in order to provide the right conditions for biogas extraction. In addition, as a by-product of this process, you can obtain bio-fertilizer [32].

The need to manage urban waste dates back to the time of the Roman Empire. They already had an environmental conscience, they worried about where their vessels and ceramics would go, and from there comes recycling. They recovered them to make other containers, used as fertilizer in agriculture or even as material for construction.

Some of the first authors to refer to biogas production were Sanghi and colleagues in 1977 [33]. This chapter pointed out the benefits of an anaerobic digester, and its generating potential for energy, where they saw it as an alternative to reduce the money invested in oil imports.

In Chile, there is great potential to generate energy with biogas from waste, not only in landfills but also in agriculture, forestry, the food industry and salmon farming. According to Ortiz, until 2017, there were 25 biodigesters nationwide, of which 10 are in the operating phase located in the Los Lagos region, the other biodigesters were in the project and start-up phase. This shows that the power generation potential has not been fully exploited [34].

When collecting information on business models applied in different parts of the world, we can find that there are five producing countries that have been able to make this product, these are Germany, Spain, Brazil, Canada and Sweden. A report from the ministry of energy of Chile mentions factors that they have in common, that is, they receive a state boost in the form of investment subsidies. In Germany, Spain and Canada the projects of biogas that generate energy for sale to the grid have a guaranteed rate. In Sweden, the use of biogas as a vehicle fuel has also been given impetus. Setting it as tax-free fuel and subsidizing the purchase of vehicles that work with biomethane [31].

It is important to mention that the process of obtaining gas from garbage is commercially viable [35]. Regarding the Chilean national regulation, there are minimum requirements for the sale of energy to the central interconnected network, which suggests that generating electricity from biogas may be a possibility of business. Jaramillo & Matthews mention that in addition to an economic benefit there is a social benefit for this type of project [36]. This allows for meeting the needs of the community, it is friendly to the environment since it does not increase the amount of carbon dioxide in the atmosphere and it constitutes a great sustainable alternative by promoting greater awareness about a more balanced relationship with nature.

A case study in Mexico makes an estimate of waste per capita of Michoacán of waste with its specific percentages for each type of material, estimation of the biogas production through the Mexican biogas model version 2.0, developed by SCS Engineers under agreement with the LMOP program of the Environmental Protection Agency of the United States (US EPA) [37]. The model generates biogas production and capture projections depending on waste management and arrangement of the sanitary landfill, in order to carry out short-term feasibility studies, medium and long term of this type of project. The description of the scenarios of this study will guide the modeling of biogas generation to be done in a larger proportion [38].

This project proved to be technically and economically feasible, the data that were required are very similar to those that will be needed in this investigation, for example, the costs of the entire project, amount of tons available at the end of the project, benefits of each ton of organic waste. The results summarized by Vera indicate that the benefit obtained from saving electricity is compared with the cost of a study that includes three important aspects: the investment cost, operation and maintenance [39]. This study shows that the scenarios studied are above the cost of a sanitary landfill, which indicates that a project with these characteristics is prefeasibility even if the biogas capture efficiency is the lowest (40%) [38].

3. Methods

As mentioned in the preceding sections, this research analyses an investment project for the creation of a biogas plant, from household organic waste in the commune of Panguipulli. This research arises from identifying a waste of the energy potential of waste in the commune. The general aspects of the project include the following stages.

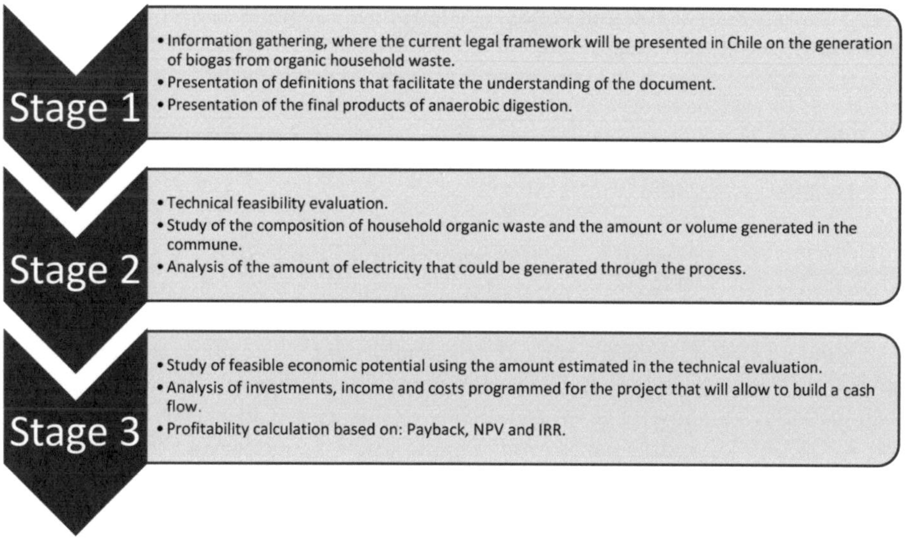

It is important to mention that the use of the previously exposed methodology allows an analysis of the technical requirements of a biogas plant, in addition, projected income will be considered and expenses to measure its potential returns.

3.1 Stage 1

3.1.1 Legal framework

All projects must comply with a minimum regulatory framework for their legal operation:

1. The Supreme Decree of Chile No. 119 of the year 2016 generated by the Ministry of Energy is related to the regulation at the construction and operational level. This decree seeks to ensure safety [40].

2. Act 20.339 of the Ministry of Mining of 1978 requires that biogas plants be registered in the electricity and fuels [41].

3. Decree 10: Regulation of boilers, autoclaves and equipment that use steam water. This decree establishes the requirements for boilers and accessories related to combustion [42].

4. Act 20.571. This law regulates the operation of electrical generation equipment. They work on the basis of non-conventional renewable energies [43].

5. Act 20.698: Promotes the expansion of the energy matrix, through non-conventional renewables sources [44].

3.1.2 End products of anaerobic digestion

From anaerobic digestion, final products are obtained such as biogas with energy-generating potential, as well as a stable biosolid that is used to improve the soil (biofertilizer or biofertilizers). This is an organic product with a high quantity of nutrients, it is not polluting and does not have pathogenic microorganisms, and finally a mixture of water and solids, the latter are obtained from the anaerobic decomposition of the substratum.

3.2 Stage 2

3.2.1 Biomass availability

In 2019 a characterization of the composition of the MSW was carried out, this showed that the total waste generated in Panguipulli is 9361 tons per year. **Figure 1** shows the MSW generated each month in 2019.

A total of 46% of the 9361 tons of household solid waste generated in Panguipulli, corresponds to household organic waste. A graph showing the composition of MSWs is shown below. According to this information, we can conclude that 4,306,060 kilograms per year of organic waste are generated domiciliary, which is equivalent to 11,961 kilograms per day.

It is suggested that organic waste be separated in homes, at the moment in which they are generated, for this the cooperation of the population of Panguipulli is needed. In this process, conscious education on the separation of waste is of vital importance in order to have biogas according to expectations. In addition, this will drive a culture towards the sustainable management of household organic waste in the commune.

Given the current pandemic situation caused by COVID-19, it will be necessary to have safeguards in the handling of organic waste [45]. This is why some authors recommend taking measures for the adequate extraction of organic waste to seek the protection of workers who are part of the collection and transport of the substrate, reducing the possibility of being infected during their workday.

3.2.1.1 Plant

Plants can produce different amounts of biogas depending on the substrate used. In this study, for all calculations, it is taken into account that the substrate is waste of organic household products, which has a biogas production capacity of 50 m^3 per ton, this substrate is among the most profitable.

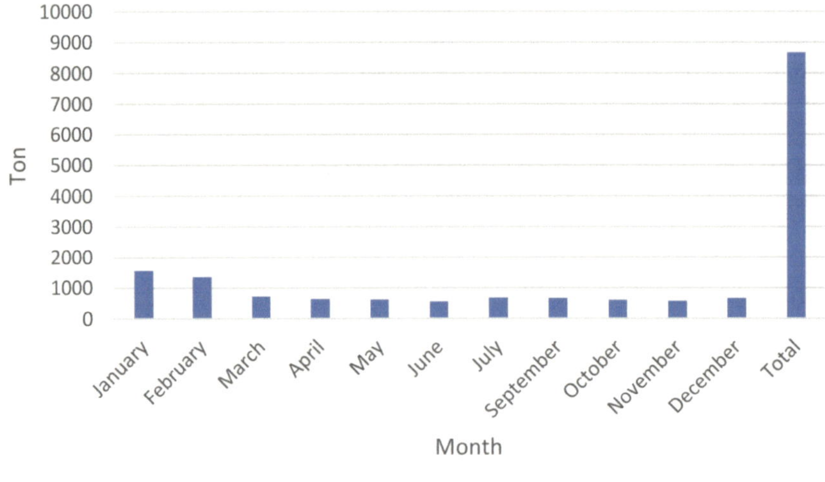

Figure 1.
Solid waste generated in 2019 in Panguipulli, Chile. Source: Department of Cleaning and Decoration of Panguipulli.

3.2.1.2 Plant operation

The biodigester generates methane from household organic waste. When methane begins to be generated, it will flow naturally to the capture point from the biodigester, from there, the electricity generator is fed, which is an engine optimized for the generation of electricity, using methane as fuel. The motor is powered by a gas pump and has a gas meter to monitor consumption.

3.2.1.3 Climatic factors

In order to implement this project, it will be important to consider climatic factors of the sector where the plant will operate, because the average climate in Panguipulli during the course of the year is between 3°C and 23°C. It may be necessary to implement a heated bioreactor and have proper insulation, this can increase some costs, since as the temperature of the biodigester increases, the speed of the growth of microorganisms increases, therefore, accelerating the digestion process obtaining a high content of methane in the biogas and conclusion obtaining profitable results.

3.2.1.4 Holding time

To generate the degradation process of organic waste or substrates time must pass, which depends directly on the temperature of the sector. For the calculations, the retention time used in biogas plants in the Los Rios Region will be taken as a reference where its substrate is also household organic waste.

Assuming its load is daily, the retention time will determine when the volume of charge needed to feed the digester is required. It is proposed to work with retention times of between 40 and 50 days and with daily loads of 10 kg per cubic meter of the digester.

3.2.1.5 Biogas generation potential estimate

The yield can be estimated according to the capacity of the biogas plant, these yield can be affected by factors such as retention time, temperature and agitation of the substrate, among others.

Table 2 shows total returns; the information used in its generation was: 1 ton is equivalent to 50 m^3 of biogas per day, 1 m^3 of biogas can generate 1.8 kilowatt-hour (kWh) per day, which delivers electrical power to the generator of 50 kW.

In addition, it is necessary to clarify that by the multiplication of the total biogas of each cubic meter per unit of KWh, the total kw per day of electrical energy production is obtained. **Table 2** with its data is shown below.

	Tons	Unit	Total biogas	Total kw/day
Biogas production per day	11,961 ton	50 m^3	598.05 m^3	
Electric power production		1.8 kwe	598.05 m^3	1076.49
Power electrical energy of the generator		50 kw		

Source: The authors.

Table 2.
Total yields from the listed biogas plant.

3.2.1.6 Ground

The land should be a flat surface, ideally, it has a sewer, to facilitate maintenance and be able to channel liquid elimination, if not, it will require trucks, and clean pits, which would increase maintenance costs. Furthermore, it is necessary that there is the availability of water. It is very important that the plant is exposed to the sun, there should not be a mound nearby to shade it. There must be good access for the trucks that will carry the raw material to enter without a problem.

3.2.1.7 Generating potential of biofertilizer

Among the by-products generated by the biogas plant, is the solid bio stable (soil improver) mentioned above in the scheme of the process of anaerobic digestion, this product can be used as a biofertilizer for soils, as it has nutrients such as potassium, phosphorus, nitrogen among others, which help to recover minerals lost in crops [46].

3.3 Stage 3: economic valuation

This section describes the aspects of the cash flow elaborated, in detail. Using the cash flow, economic indicators were obtained that allow evaluation of the profitability of the project. The calculations are presented in Appendix 1, with their respective NPV, IRR and Payback. These tools are the most suitable in this investigation as it allows the calculation of the time in which the initial investment will be recovered to be made more precise.

3.3.1 Income from the sale of electricity

It is one of the main incomes, which corresponds to 60% of the energy produced, it will be injected into the distribution network that under Law 20,571 has entered into force since 2014. For the calculations, we express prices and costs in CLP, 830 CLP is equivalent to 1 US dollar. A price of 385 CLP/kW was estimated, in addition the plant has an electric power of the generator of 50 kW, which generates 31,200 kW-month, the plant will produce electricity 24 hours a day for 6 days a week, taking a total of 4 days a month for maintenance. The income per sale will be constant over time. The above delivers a total annual income of 86.486.400 CLP. This information was collected from the data historical prices of the node near Panguipulli [47].

3.3.2 Income from energy savings in self-consumption

Energy generated by the biogas plant allows it to pay for the monthly energy supply which corresponds to 40% of all electrical energy produced. For this, the amount of kWh saved annually was valued by installing the biogas plant and the economic savings incurred were estimated. Therefore, an annual saving in electrical energy of 57.657.600 CLP is obtained.

3.3.3 Income from sale of biofertilizer

It was neither possible to find the value of the fertilizers that are used in the market nor the sale value of biofertilizers generated by biogas plants, since these depend on the chemical compositions. Therefore, to determine in some way the income of the biofertilizer, the price for sale at 44.8 CLP per kg was used and the percentage of recovery of organic matter for the generation of biofertilizer of 30% with respect to the initial organic matter. These data were recovered from a study of a biogas plant using grape marc as substrate [46]. Income from the sale of biofertilizer is equal to 58.060.800 CLP annually and will remain constant over time. The costs associated with this project are investment costs, operating and maintenance costs and costs for investments in intangibles.

3.3.4 Investment costs

It can be seen in the following table that the total cost of the investment required by the biogas plant amounts to 540.000.000 CLP (810 CLP are approximately 1 US dollar). The estimation percentages of the factors influencing the project were taken from Garay García thesis (see **Table 3**) [48].

3.3.5 Operation and maintenance costs

These costs are associated with the substrate (since currently the substrate is not used in anything, it does not have a cost or price), maintenance, waste disposal, costs of operating inputs and personnel costs. Total operating costs amount to 53.824.457 CLP yearly.

Regarding the personnel requirements, it is considered for the calculations that it is necessary to work with five people for the operation of the plant, where two people are technicians and work full-time and the others are full-time assistants. Estimates of personnel cost are 27.720.000 CLP per year.

Regarding the costs of inputs, water for the tributary of the digester is necessary, for the calculations of water used for loads of the tributaries, it is estimated 714.457 CLP yearly [49].

Another cost to consider is the maintenance and repairs of the equipment, this will be calculated based on percentages of the total investment cost. Total maintenance costs are equivalent to 25.390.000 CLP per year (see **Table 4**) [48].

3.3.6 Cost of investment in intangibles

This cost includes patents to function in a legal form, contracts, insurance for damage to equipment or motors, pumps, agitators, among others. In addition, it is recommended to take out insurance in case of earthquakes or other situations that may damage the investment. The cost associated with intangibles varies between 0.8% and 1% of total investments [46]. The total investment amounts to 540.000.000 CLP and must be considered in year 0.

Biodigester and cogeneration engine	270.000.000
Installation (30%)	81.000.000
Engineering (15%)	40.500.000
Start-up (15%)	40.500.000
Civil works (10%)	27.000.000
Electrical equipment, pipes and insulating materials (15%)	40.500.000
Contingencies (15%)	40.500.000
Net investment cost	**540.000.000**

Source: [47].

Table 3.
Net investment cost in CLP.

Civil works (1%)	2.170.000
Equipment (4%)	1.620.000
Biodigester and cogenerator (8%)	21.600.000
Total cost of maintenance	25.390.000
Source: [47].	

Table 4.
Total cost of maintenance in CLP.

3.3.7 Land rental

Considering that this project could be executed by a municipality, as well as a private company, an estimated market rental value for urbanized land in the surroundings of the city of Panguipulli is equivalent to 3.600.000 CLP per year. The value is constant over time and is exempt from tax in accordance with the provisions of Exempt Resolution No. 300, of 1970, revalidated in accordance with instructions contained in Decree No. 111 of 1975 of Chile [50].

Year	Biodigester and cogeneration engine 270.000.000	Civil works 27.000.000	Equipment 40.500.000	Intangible 5.400.000	Total depreciation
0					-
1	27.000.000	1.350.000	4.050.000	1.800.000	34.200.000
2	7.000.000	.350.000	4.050.000	1.800.000	34.200.000
3	27.000.000	1.350.000	4.050.000	1.800.000	34.200.000
4	27.000.000	1.350.000	4.050.000		32.400.000
5	27.000.000	1.350.000	4.050.000		32.400.000
6	27.000.000	1.350.000	4.050.000		32.400.000
7	27.000.000	1.350.000	4.050.000		32.400.000
8	27.000.000	1.350.000	4.050.000		32.400.000
9	27.000.000	1.350.000	4.050.000		32.400.000
10	27.000.000	1.350.000	4.050.000		32.400.000
11		1.350.000			1.350.000
12		1.350.000			1.350.000
13		1.350.000			1.350.000
14		1.350.000			1.350.000
15		1.350.000			1.350.000
16		1.350.000			1.350.000
17		1.350.000			1.350.000
18		1.350.000			1.350.000
19		1.350.000			1.350.000
20		1.350.000			1.350.000
Source: [47].					

Table 5.
Depreciation per asset individually in CLP.

3.3.8 Working capital

The project will generate income from its start-up by the sale of electrical energy and biofertilizer, a working capital will be estimated that allows it to cover the first 3 months of operation, this includes rent, and operating cost and maintenance. The working capital is equivalent to 163.456.113 CLP annually.

3.3.9 Depreciation

Depreciation corresponds to the decrease in the value of assets due to their use or deterioration. Depreciation in this project was estimated with a normal useful life. Below is the depreciation of each asset individually, the number of years of useful life was extracted from the SII website (see **Table 5**) [50].

4. Discussion

The technical evaluation of the biogas project from household organic waste for the production of electrical energy, self-consumption and sale of bio fertilizers, projects that the process is technically feasible mainly due to the fact of the substrate nowadays. It is a problem with high costs for the municipality, and for this project, it is free raw material.

For the start-up of the project, it is important to consider that the costs of the investment evaluated are mainly concentrated in the biodigester and the cogeneration, being 50% of the investment. This indicates that it is very important to know the real cost of this equipment. It is recommended to obtain quotes from several companies, in addition to calculating the dimensions, since this could affect the cost of the investment which would affect the profitability of the project.

For the execution of the project, the variability in time of electricity prices is a consideration. In this study, it was considered that energy production would be sold to the central interconnected system, but there is also another option that was not estimated since it is currently not very feasible. The sale of energy directly to companies, could generate contracts for long periods, but the investment of the installation of wiring and other costs, in the city of Panguipulli there does not currently exist a large company that could be a potential client.

It is important to recognize that a weakness of the project is its high investment cost, which means an entry barrier to the energy and fertilizer market. As was commented previously, the waste for another entity means an expense, but seeing it from this perspective that it is a potential income generator, in addition to being an environmentally friendly process, it provokes an acceptance by the surrounding communities and could be considered in the municipality plan.

The cash flow indicates that the project under the conditions defined in the technical and economic evaluation is profitable according to the economic indicator of net present value. It amounts to 214.099.637 CLP, an IRR of 15% and a recovery period of 6 years.

This is because the sale price of bio fertilizer is high, capable of absorbing almost all annual expenses, a high percentage of sales can be estimated, since the market for fertilizers in this area is great. The information previously presented allows us to answer the hypotheses; where the first two are accepted, the construction of a biogas plant in the Panguipulli commune is economically profitable and the volume of household organic waste in the Panguipulli commune makes it possible to construct a biogas plant, while the third hypothesis is refuted. The investment in the

NPV	**214.099.637**
IRR	15%
Payback	6 years

Table 6.
Results of economic indicators in CLP.

construction and implementation of a biogas plant in the commune of Panguipulli is recovered in 4 years.

The evidence of the results of the indicators economic showed that investment in the construction and implementation of a plant of biogas in the commune of Panguipulli is recovered in 6 years (see **Table 6**).

It is recommended that to reduce the risk of the project, the given climatic factor be considered, which is the greatest limitation for the development of biodigestion projects. This inconvenience can be reduced by implementing complementary heating to the biodigester and implementing proper insolation. In this way, by increasing biogas production, consequently, the generation power of kW will increase the profitability of the system. It is also suggested to include information on the location of the project, this will allow knowledge of environmental conditions, wind speed and direction.

The technical and economic evaluation of a biogas plant from household organic waste allowed a visualization of the economic profitability, points to consider and difficulties that will allow clarification of the situation to potential investors.

5. Conclusion

Currently, there is excessive growth in waste production at a worldwide level that leads to the search for new solutions that allow the reuse of waste in a sustainable way over time and friendly to the environment, within these options is biogas, that by means of a biodigester offers advantages for the waste treatment which generates a gaseous fuel, which can be used to generate electrical energy. It also generates a quality biofertilizer and with this, it is possible to reduce the environmental damage caused by accumulating this substrate in a sanitary landfill.

When analyzing the composition of the waste, it was calculated that 12 tons of household organic waste per day is generated in the Panguipulli commune. This allowed the size of an appropriate biodigester to store 40 days of retention, and thus generate 600 m^3 of biogas per day, which provides electrical power of the generator of 50 kW that allows a generation per year of 374,400 kW-year. Thanks to this, it can self-consume energy electricity and sell the rest to the central interconnected system.

This research has several characteristics that position it with a potential for biogas production, these are; availability of substrate use, geographic availability of the substrate, stable prices and costs and projected in time and finally to create a project that minimizes environmental impact.

Finally, the economic evaluation obtained a net present value (NPV) of the project evaluated to 15 years of 214,099,637 and an internal rate of return (IRR) of 15% to a real discount rate of 10%. The investment payback period is 6 years.

Acknowledgements

This chapter was written based on Cristina Salazar's commercial engineering thesis at the Universidad Austral de Chile, supervised by Professor Jean Pierre Doussoulin.

Appendix 1: Cash flow in millions of Chilean pesos (CLP).

CASH FLOW	0	Year 1	Year 2	Year 3	Year 4	Year 5	Year 6	Year 7	Year 8	Year 9	Year 10	Year 11	Year 12	Year 13	Year 14	Year 15
Income from the sale of biofertilizer	58.1	58.1	58.1	58.1	58.1	58.1	58.1	58.1	58.1	58.1	58.1	58.1	58.1	58.1	58.1	58.1
Income from the sale of electricity	86.5	86.5	86.5	86.5	86.5	86.5	86.5	86.5	86.5	86.5	86.5	86.5	86.5	86.5	86.5	86.5
Electric energy saving	57.7	57.7	57.7	57.7	57.7	57.7	57.7	57.7	57.7	57.7	57.7	57.7	57.7	57.7	57.7	57.7
Total revenues	202.2	202.2	202.2	202.2	202.2	202.2	202.2	202.2	202.2	202.2	202.2	202.2	202.2	202.2	202.2	202.2
Operation and maintenance costs		53.8	53.8	53.8	53.8	53.8	53.8	53.8	53.8	53.8	53.8	53.8	53.8	53.8	53.8	53.8
Land lease		3.6	3.6	3.6	3.6	3.6	3.6	3.6	3.6	3.6	3.6	3.6	3.6	3.6	3.6	3.6
Total expenses	0	57.4	57.4	57.4	57.4	57.4	57.4	57.4	57.4	57.4	57.4	57.4	57.4	57.4	57.4	57.4
Gross profit	0	144.8	144.8	144.8	144.8	144.8	144.8	144.8	144.8	144.8	144.8	144.8	144.8	144.8	144.8	144.8
Tax	0	−39.1	−39.1	−39.1	−39.1	-39.1	39.1	39.1	39.1	39.1	39.1	39.1	39.1	39.1	39.1	39.1
Profit after tax	0	105.7	105.7	105.7	105.7	105.7	105.7	105.7	105.7	105.7	105.7	105.7	105.7	105.7	105.7	105.7
Depreciation	34.2	34.2	34.2	34.2	34.2	34.2	34.2	34.2	34.2	34.2	1.4	1.4	1.4	1.4	1.4	34.2
Net Investments	540.0															
Investment VAT	102.6															
VAT recovery		102.6														
Working capital	163.5															
Recovery of working capital																163.3
Investment in intangibles	5.4															
capital flow	−811.5	139.9	139.9	139.9	139.9	139.9	139.9	139.9	139.9	139.9	139.9	107.0	107.0	107.0	107.0	107.0
Cumulative Cash Flow	−811.5	−671.6	−531.7	−391.8	−251.9	−112.0	27.9	167.8	307.7	447.6	587.4	694.5	801.5	908.6	1.015.6	1.122.6

Author details

Jean Pierre Doussoulin[1,2*] and Cristina Salazar Molina[3]

1 Universidad Austral de Chile, Economic Institute, Chile

2 Université Gustave Eiffel, Research Team on the Use of Panel Data in Economics, France

3 Universidad Austral de Chile, Faculty of Economics and Administrative Sciences, Chile

*Address all correspondence to: jean.doussoulin@uach.cl; jean-pierre.doussoulin@u-pem.fr

References

[1] Doussoulin JP. A paradigm of the circular economy: The end of cheap nature? Energy, Ecology and Environment. 2020;**5**(5):359-368

[2] Otterpohl R, Albold A, Oldenburg M. Source control in urban sanitation and waste management: Ten systems with reuse of resources. Water Science and Technology. 1999;**39**(5):153-160

[3] Rivas-Solano O, Faith-Vargas M, Guillén-Watson R. Biodigestores: factores químicos, físicos y biológicos relacionados con su productividad. Revista Tecnología en Marcha. 2010; **23**(1):39

[4] Garfí M, Ferrer-Martí L, Velo E, Ferrer I. Evaluating benefits of low-cost household digesters for rural Andean communities. Renewable and Sustainable Energy Reviews. 2012;**16**(1): 575-581

[5] Mayer A, Haas W, Wiedenhofer D, Krausmann F, Nuss P, Blengini GA. Measuring Progress towards a circular economy: A monitoring framework for economy-wide material loop closing in the EU28. Journal of Industrial Ecology. 2019;**23**(1):62-76

[6] Surendra KC, Takara D, Hashimoto AG, Khanal SK. Biogas as a sustainable energy source for developing countries: Opportunities and challenges. Renewable and Sustainable Energy Reviews. 2014;**31**:846-859

[7] Bond T, Templeton MR. History and future of domestic biogas plants in the developing world. Energy for Sustainable development. 2011;**15**(4):347-354

[8] Mamo TZ, Dutta A, Jabasingh SA. Start-up of a pilot scale anaerobic reactor for the biogas production from the pineapple processing industries of Belgium. Renewable Energy. 2019;**134**: 241-246

[9] Salimbeni A. Sweet sorghum biogas plant in temperate regions (Belgium)-demonstration plant for biogas and high value biofertilizer production. In: En: Proceedings of the 21st European Biomass Conference and Exhibition. Copenhagen, Denmark; 2013. pp. 3-7. Available from: https://scholar.google.com/scholar?hl=es&as_sdt=0%2C5&q=Sweet+sorghum+biogas+plant+in+temperate+regions+%28Belgium%29+demonstration+plant+for+biogas+and+high+48+value+biofertilizer+production&btnG=

[10] Lamnatou C, Nicolaï R, Chemisana D, Cristofari C, Cancellieri D. Biogas production by means of an anaerobic-digestion plant in France: LCA of greenhouse-gas emissions and other environmental indicators. Science of the total environment. 2019;**670**:1226-1239

[11] Almansour E, Bonnet J-F, Heredia M. Potentiel de production de biogaz à partir de résidus agricoles ou de cultures dédiées en France. Sciences Eaux Territoires. 2011;**1**:64-72

[12] Martinez-Perez N, Cherryman SJ, Premier GC, Dinsdale RM, Hawkes DL, Hawkes FR, et al. The potential for hydrogen-enriched biogas production from crops: Scenarios in the UK. Biomass and Bioenergy. 2007;**31**(2–3):95-104

[13] Patterson T, Esteves S, Dinsdale R, Guwy A. An evaluation of the policy and techno-economic factors affecting the potential for biogas upgrading for transport fuel use in the UK. Energy Policy. 2011;**39**(3):1806-1816

[14] Morero B, Vicentin R, Campanella EA. Assessment of biogas production in Argentina from co-digestion of sludge and municipal solid waste. Waste management. 2017;**61**:195-205

[15] Fitzsimons RE, Laurino CN, Vallejos RH. Estimation of potential

biomass resource and biogas production from aquatic plants in Argentina. Energy. 1982;7(8):681-687

[16] Nascimento MCB, Freire EP, de Dantas F AS, Giansante MB. Estado da arte dos aterros de resíduos sólidos urbanos que aproveitam o biogás para geração de energia elétrica e biometano no Brasil. Engenharia Sanitaria e Ambiental 2019;**24**:143-155

[17] Barros RM, Tiago Filho GL, Santos AHM, Ferreira CH, Pieroni MF, Moura JS, et al. A potential of the biogas generating and energy recovering from municipal solid waste. Renewable Energy Focus. 2018;**25**:4-16

[18] Demuynck M, Naveau H, Nyns E-J. Biomethanation and biogas plants in the USA—A European view. Trends in biotechnology. 1984;**2**(1):2-5

[19] Dale BE, Bozzetto S, Couturier C, Fabbri C, Hilbert JA, Ong R, et al. The potential for expanding sustainable biogas production and some possible impacts in specific countries. Biofuels, Bioproducts and Biorefining. 2020; **14**(6):1335-1347

[20] Okudoh V, Trois C, Workneh T, Schmidt S. The potential of cassava biomass and applicable technologies for sustainable biogas production in South Africa: A review. Renewable and Sustainable Energy Reviews. 2014;**39**: 1035-1052

[21] Uhunamure SE, Nethengwe NS, Tinarwo D. Correlating the factors influencing household decisions on adoption and utilisation of biogas technology in South Africa. Renewable and Sustainable Energy Reviews. 2019; **107**:264-273

[22] Ncube A, Cocker J, Ellis D, Fiorentino G. Biogas from source separated organic waste within a circular and life cycle perspective. A case study in Ontario, Canada. Environmental and Sustainability Indicators. 2021;**11**:100134

[23] Paul S, Dutta A, Defersha F, Dubey B. Municipal food waste to biomethane and biofertilizer: A circular economy concept. Waste and Biomass Valorization. 2018;**9**(4):601-611

[24] Chen Y, Yang G, Sweeney S, Feng Y. Household biogas use in rural China: A study of opportunities and constraints. Renewable and Sustainable Energy Reviews. 2010;**14**(1):545-549

[25] Qu W, Tu Q, Bluemling B. Which factors are effective for farmers' biogas use?–evidence from a large-scale survey in China. Energy Policy. 2013;**63**:26-33

[26] Casas R, Casas F, Bustos J. Design of profitable networks of biogas plants in Chile. SN Applied Sciences. 2021;**3**(6): 635

[27] Gaete-Morales C, Gallego-Schmid A, Stamford L, Azapagic A. Assessing the environmental sustainability of electricity generation in Chile. Science of The Total Environment. 2018;**636**: 1155-1170

[28] González Martínez T, Bräutigam K-R, Seifert H. The potential of a sustainable municipal waste management system for Santiago de Chile, including energy production from waste. Energy, Sustainability and Society. 2012;**2**(1):24

[29] Mao G, Zou H, Chen G, Du H, Zuo J. Past, current and future of biomass energy research: A bibliometric analysis. Renewable and Sustainable Energy Reviews. 2015;**52**:1823-1833

[30] Parra-Orobio BA, Nieto-Mendoza M, Rivera-Henao D, Manyoma-Velásquez PC, Torres-Lozada P. Selection of inocula conditioning methodologies for the anaerobic digestion of food waste. Revista

Facultad de Ingeniería Universidad de Antioquia. 2019;**92**:9-18

[31] Gamma Engineers SA. Modelos de negocio que rentabilicen aplicaciones de biogás en Chile y su fomento. Ministerio de energía. 2011. Available at https://www.aproval.cl/manejador/resources/informefinalmodelosbiogasabril2011.pdf

[32] Grass Puga BD. Evaluación y diseño para la implementación de una planta de biogas a partir de residuos orgánicos agroindustriales en la Región Metropolitana. 2013

[33] Sanghi AK, Day D. A cost-benefit analysis of biogas production in rural India: Some policy issues. In: En: Agriculture and Energy. Elsevier; 1977. pp. 607-624. Available from: https://www.sciencedirect.com/science/article/pii/B9780124542501500496

[34] Ortiz S. Metodología práctica de evaluación técnico-económico-ambiental y gestión de utilización de desechos orgánicos en industria de ganado para la producción de energía eléctrica. 2017

[35] Scheutz C, Kjeldsen P, Bogner JE, De Visscher A, Gebert J, Hilger HA, et al. Microbial methane oxidation processes and technologies for mitigation of landfill gas emissions. Waste Management & Research. 2009;**27**(5):409-455

[36] Jaramillo P, Matthews HS. Landfill-gas-to-energy projects: Analysis of net private and social benefits. ACS Publications. 2005. Available from: https://pubs.acs.org/doi/abs/10.1021/es050633j

[37] Aguilar-Virgen Q, Taboada-González PA, Ojeda-Benítez S. Modelo mexicano para la estimación de la generación de biogás. Ingeniería. 2011;**15**(1):37-45

[38] Vera-Romero I, Estrada-Jaramillo M, Martínez-Reyes J, Ortiz-Soriano A. Potencial de generación de biogás y energía eléctrica. Parte II: residuos sólidos urbanos. Ingeniería, investigación y tecnología. 2015;**16**(3):471-478

[39] Chong TL, Matsufuji Y, Hassan MN. Implementation of the semi-aerobic landfill system (Fukuoka method) in developing countries: A Malaysia cost analysis. Waste Management. 2005;**25**(7):702-711

[40] Ministry of Energy of Chile. Supreme Decree N°119. 2016. Avaliable at https://www.bcn.cl/leychile/navegar?idNorma=1099847&idVersion=2017-07-01&idParte=

[41] Ministry of Mining of Chile. Act 20.339. 1978. Available at https://www.cne.cl/archivos_bajar/LEY_20.339.pdf

[42] Superintendency of Health of Chile. Act 10: Reglamento de calderas, autoclaves y equipos que utilizan vapor de agua. 2013. Available at http://www.supersalud.gob.cl/observatorio/671/w3-article-8910.html

[43] Ministry of Energy of Chile. Act 20,571. 2018. Available at https://www.bcn.cl/leychile/navegar?idNorma=1038211

[44] Ministry of Energy of Chile. Act 20.698. 2013. Available at https://www.bcn.cl/leychile/navegar?idNorma=1055402

[45] Doussoulin JP, Mougenot B. The Economic, Climate Change and Public Health Edges of the Geopolitics of COVID-19: An Exploratory Bibliometric Analysis. En: Biotechnology to Combat COVID-19. London, UK: IntechOpen; 2021

[46] Fernández C. Analisis técnico economico para instalación planta de

biogás utilizando residuos vitivinícolas para generación de energía en la región del Maule. Universidad técnica federico Santa Maria, Industries department. Available at https://repositorio.usm.cl/bitstream/handle/11673/24608/3560900257247UTFSM.pdf?sequence=1&isAllowed=y; 2018

[47] National Energy Commission of Chile. Data of historical prices and investment in energy sector of Chile. 2020. Available at https://www.cne.cl/en/estudios/electricidad/

[48] Garay García OA. Análisis técnico-económico de una planta de generación eléctrica en base a biogás. 2011. Available at https://repositorio.uchile.cl/handle/2250/104119

[49] Fernandez C. Análisis técnico económico para instalación de planta de biogás utilizando residuos vitivinícolas para generación de energía en la región del Maule. 2018; Disponible en: https://repositorio.usm.cl/handle/11673/24608

[50] Tax Bureau of Chile. Nueva tabla de vida útil de los bienes físicos del activo inmovilizado. 2020. Available at https://www.sii.cl/pagina/valores/bienes/tabla_vida_enero.htm

Chapter 7

Case Studies in Biogas Production from Different Substrates

Adrian Eugen Cioabla and Francisc Popescu

Abstract

The present paper involves applicative research in the field of biogas production with the accent on small laboratory scale installations built for biogas production, preliminary testing of substrate for biogas production and combustion applications for biogas-like mixtures. The interconnected aspect of the presented material involves cumulative expertise in multidisciplinary fields of interest and continuous development of possibilities to determine the energetic potential of substrates subjected to biodegradable fermentation conversion for further applications. The research analyzed the combustion behavior of biogas with different methane/carbon dioxide ratio without and in the presence of specific catalysts. Also, laboratory analysis on biomass substrates for determining their physical and chemical potential for different applications was performed. The main conclusions are drawn revolve around the untapped potential of the different types of biomasses that are not commonly used in the production of renewable energy carriers, like biogas, and also the potential use of residual biomass in combustion processes for an enclosed life cycle from cradle to the grave. The study involving the use of catalysts in biogas combustion processes present possible solutions which can be developed and implemented for increasing the combustion quality by using relatively cost-effective materials for the production of catalytic materials.

Keywords: biomass, biogas, anaerobic digestion, renewable energy, combustion

1. Introduction

Renewable energy sources have become a milestone of the next decades for European Union member states, due to tough deadline targets to reduce their energy production dependency on fossil fuels. In June 2021, thru its new European Climate Law, the EU targets a reduction of greenhouse gases emission of at least 55% by 2030, compared to the 1990 level, and a net-zero greenhouse gas emissions in the EU by 2015. On top of this ambitious target, the EU is pushing its member states to complete climate neutrality by 2050. In this frame, the new 2030 greenhouse gases emission reduction target for Romania is 12.5% (in respect to 2005 inventory), a target that Romania has already passed in several chapters, with a total reduction in 2019 of 65% compared to 1990 emissions inventory, being the member state that achieved the highest greenhouse gases emissions reduction in the EU, along with Lithuania. However, for a specific chapter, the "transportation, buildings and agricultures sector" Romania has a new reduction target of 12.7% by 2030 (compared to 2005 national GHG inventory) [1].

Having in mind that for Romania most of the industrial sectors have or are close to reach their targets for 2030 and the main 2 sectors that still must reduce with 12.7% of their greenhouse gases emissions by 2030 are transportation and agriculture are clear that the focus should be on developing the renewable sources that can have a significant impact on both sectors. The renewable energy sector had a fast development in the past 2 decades, however, the energy production from biomass/biogas did not increase as other sources. For example, in 2019 in Romania the installed renewable energy installed capacities from biomass/biogas was of 124.16 MW while the photovoltaic installed capacities were 1358.43 MW and wind farms at 2960.64 MW. With this data on mind it becomes quite clear that Romania's focus in the next period should be on biomass/biogas production facilities development as not only will reduce dependency on methane imports but will contribute to reaching greenhouse gasses emissions from the agriculture sector.

The production of biogas from biomass substrates thru anaerobic digestion is well known since antiquity, the technology being constantly developed but due to environmental impact thru pollutants developed (solid, gas and liquids) faces continuous challenges, with continuous scientific efforts in research for innovative materials to be used in biogas production from biomass and urban waste waters [2].

In the Romanian case and any other country with significant agricultural areas and also a large number of urban agglomerations, the potential sources for biogas production thru anaerobe digestion can be classified in four main groups:

- Organic wastes from agricultural/zootechnie – with all potential biodegradable materials;
- Agriculture and food waste, such as manure and wastes from treatment plants in form of organic-rich sludge;
- Municipal wastes in form of fermentable fractions;
- Sludge from municipal wastewater treatment plants [3].

In terms of biogas production from any of the sources classified above, the most critical parameter in obtaining the best CH_4/CO_2 ratio after anaerobic digestion is the substrate composition, as today's substrates are mostly formed for co-digestion of a minimum two waste materials. Depending on the substrate composition the anaerobic reaction temperature will have a different effect on variation of pH, volatile fatty acids, total solid degradation and ammonia/nitrogen ratio that would affect the stability of the digestion process [4].

In the EU country, all biofuels (solid, liquid or gaseous) have a subsidy from EU public budges and accordingly the production of biofuels is subject to sustainability criteria under the Renewable Energy Directive [5]. The Directive introduces significant restrains in the production of raw agricultural materials for energy use, mainly to protect primary forests and biodiversity. In respect of the Directive, the Romanian focus for developing a sustainable biofuel (with emphasis on biogas production for energy purposes) industry should be on agricultural and urban wastes.

2. Applied research on case study

Currently, the focus is to work on small-scale installations for testing in a controlled environment the potential for different materials and the possibility to develop new bioreactors for further use in anaerobic fermentation processes.

The present chapter will highlight a part of the research conducted so far, covering three main parts:

- Testing in firing processes of biogas and biogas – like mixtures – for this study, the used biogas recipe contained methane and carbon dioxide in known volume participation, initial tests were performed without catalysts and further determinations were carried on using different types of laboratories determined catalysts to determine their potential influence during the combustion process;

- Laboratory analysis on biomass substrates for determining their physical and chemical potential for different applications – the analysis for chosen biomass was performed by the European Standards for solid biofuels (EN ISO 18134 – Solid biofuels – Determination of moisture content – Oven dry method (3), EN ISO 18122 – Solid biofuels – Determination of ash content; EN 14918 – Solid biofuels –Determination of calorific value; EN ISO 16948 – Solid biofuels – Determination of total content of carbon, hydrogen and nitrogen; EN ISO 16994 – Solid biofuels – Determination of total content of sulfur and chlorine; EN ISO 18123 – Solid biofuels – Determination of the content of volatile matter; CEN/TS 15370 – Determination of ash melting behavior);

- Laboratory studies for biogas production and system development in terms of parameter monitoring and initial inputs for new different materials used for anaerobic digestion processes – the test rigs for laboratory determinations were developed in house and the main testing conditions involved using a known temperature regime (mesophilic or thermophilic), the existence or absence for materials homogenization, and continuous measuring of pH, and volume participation of methane, carbon dioxide and hydrogen sulphide in the produced biogas.

2.1 Testing in combustion processes of biogas: like mixtures

Biogas like mixtures represent, in our case, mixtures containing 70–75% methane and around 25–30% carbon dioxide, concentrations by volume, to assess the energetic potential for this type of materials, by comparison with real biogas testing, which contains also other elements, like hydrogen sulfide (the main corrosive and toxic component in biogas), ammonia, water and other impurities from the process of anaerobic digestion.

First, we will present in short, some determinations relative to biogas determinations. Those determinations were made for determining the biogas potential in firing processes and were carried out in situ, by using pilot patented installations, but for our discussions, the test rigs will not be presented, only the part needed for the firing testing.

The tests were carried out at a location for an industrial partner for Politehnica University, and the produced biogas came from anaerobic digestion of municipal residues.

The figure below presents the test rig developed for firing tests (**Figure 1**).

As it can be observed above, from left to right there are the following components: biogas pipe, connected with the system for pressure control and measuring, the burner (in yellow) and the entrance to the firing chamber, where the tests were carried out. At the end of the chamber, there were measured the flue gas and the temperatures were determined at specific points on the outside wall of the testing chamber. The next images will present some results for the measurements of the flue gas. The equipment used in this regard was TESTO 350XL and DELTA 1600 S IV gas analyzers.

Figure 1.
Elements of the test rig.

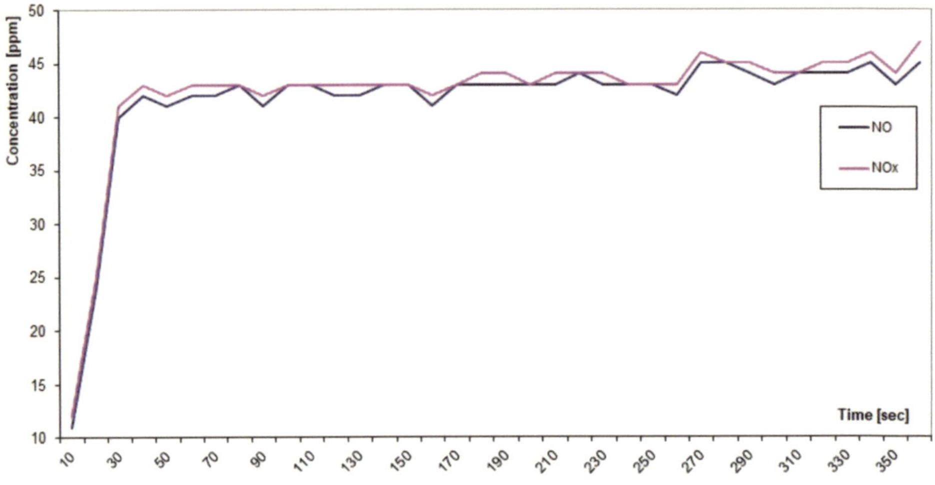

Figure 2.
NO, NOx concentration evolution in time.

Figure 2 presents the time variation of NO and NOx during the combustion process of biogas containing around 75% methane and 24% carbon dioxide. The produced biogas was without a filtering system. From the presented graphic, it can be determined that the nitrogen oxides concentration is very low (ppm values), at around 40–43 ppm, which represents very good results in this context.

The used burner had a constructive air-cooled system of the flue gas and by this method, combined with a relatively high rate of combustion, the resulting NOx emission was very low, which represents a positive aspect in this context (**Figure 3**).

Carbon monoxide is one of the most dangerous flue gasses in high quantities and it needs relatively high temperatures and safe firing conditions to be present in low concentrations. The maximum values for CO concentration during the process are also very low, at around 35 ppm, a very good indicator for a relatively complete

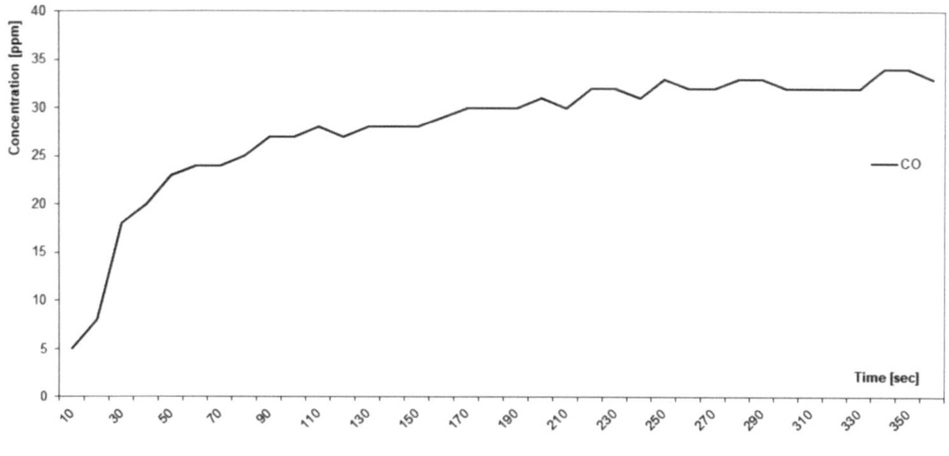

Figure 3.
CO concentration evolution in time.

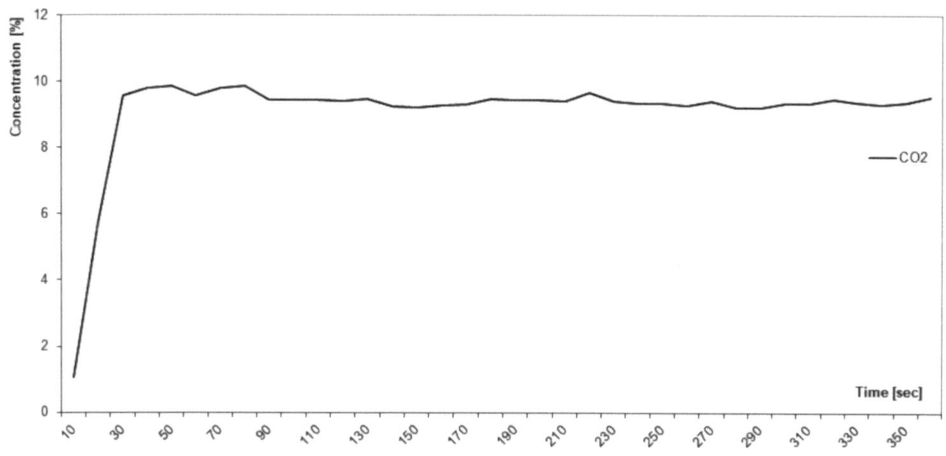

Figure 4.
CO_2 concentration evolution in time.

firing process. The obtained values used by parallel measurement with all the other existent flue gas, indicated the low volume presence of CO which is a positive argument for a very safe firing process.

As observed in **Figure 4**, CO_2 concentration maximum values were at around 9–10%. It is important to have in mind the fact that those values started at around 25–30% by volume before the firing process, which indicates also that the firing parameters were efficient, even if the overall CO_2 did not burn (because of its inert nature to firing reaction). Next, there are going to be presented tests made together with collaborators from Serbia, the Mechanical Engineering Faculty in Belgrade.

The tests were carried out in the presence and absence of catalysts to observe their influence over the firing parameters and also the flue gas was analyzed with the help of a Horiba gas analyzer coupled with a special developed system used for data collection and registration, containing temperature and pressure sensors, and data control and storage equipment (**Figure 5**).

The used catalysts were $ZnAl_2O_4$, $CoAl_2O_4$ and$ZnCr_2O_4$. The obtained pellets were inserted in a metal matrix for protection purposes and the firing chamber was prepared for preliminary tests.

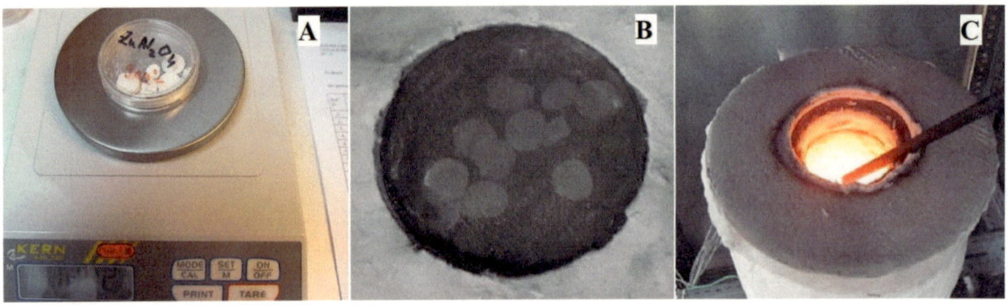

Figure 5.
Preparation of $ZnAl_2O_4$ catalysts: A – Weighting; B – Insertion in the metal matrix; C – Initial testing with and without catalysts.

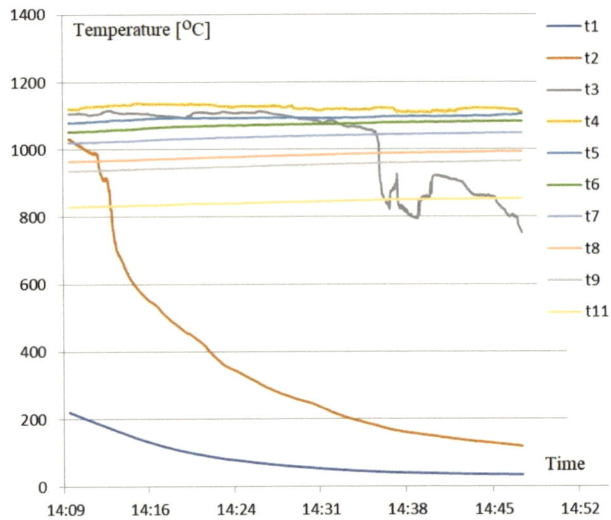

Figure 6.
Temperature values inside the combustion chamber, without catalyst presence.

Before recorded measurements, there were made some preliminary trials to calibrate all the necessary equipment and sensors on the used firing chamber. The next part will present in summary just a small part of the determinations performed inside the firing chamber, with and without the existence of catalysts (**Figure 6** and **Table 1**).

From the gathered data, it was determined that at the base of the reactor, the maximum temperature reached was around 1058°C, with very low gaseous emissions (**Figure 7** and **Table 2**).

By comparison with the first scenario, it can be observed that the maximum temperature reached at the solid phase (the base of the reactor) is around 1097°C, slightly higher than for the process without catalyst, but it was observed an increase of CO concentration, which indicated an incomplete combustion process. The main indicator of an increased CO is usually the area with high temperatures. This aspect combined with an ineffective air/fuel ratio can have as a main result the higher CO concentration, at least this is the author's present explication to this phenomenon.

The only catalyst presented in this study was $ZnCr_2O_4$, because for the other used catalysts, there was no visible influence over the firing parameters overall, this meaning they had a very limited impact in this testing scenario. Of course, further testing is to be made available to determine possible applications for the used catalysts and to test new ones for better results over impact during combustion processes.

Excess air				[–]	2.0				
Burner power				kW	1.5				
Fuel ratio CH_4:CO_2				[–]	80.4:19.6				
Fuel lower heating value				[kJ/m^3]	28847.52				
Fuel consumption				[m^3/h]	0.1872				
Air consumption				[kg/h]	3.7041				
Burner pebble bed diameter				[mm]	13				
Burner pebble bed height				[mm]	253				
Burner pebble bed porosity				[–]	44.94				
Flame arrestor bed diameter				[mm]	6				
Flame arrestor bed height				[mm]	107.5				
Temperature distribution within the porous burner – gaseous phase									
−0.1	0	0.11	0.22	0.33	0.44	0.55	0.66	0.77	0.99
t_1	t_2	t_3	t_4	t_5	t_6	t_7	t_8	t_9	t_{11}
80.5	354	1022.3	1123.8	1093.5	1072.5	1036.5	980.4	951.9	843.2
Temperature distribution within the porous burner – solid phase									
0		0.44		0.66		0.88			
t_{s1}		t_{s3}		t_{s4}		t_{s5}			
784.3		1058.2		1012.6		943.0			
Flue gas analysis at the exit of the burner									
O_2 [%]		CO_2 [%]		NOx [ppm]		CO [ppm]			
11.10		6.63		1		28			

Table 1.
Testing without catalyst.

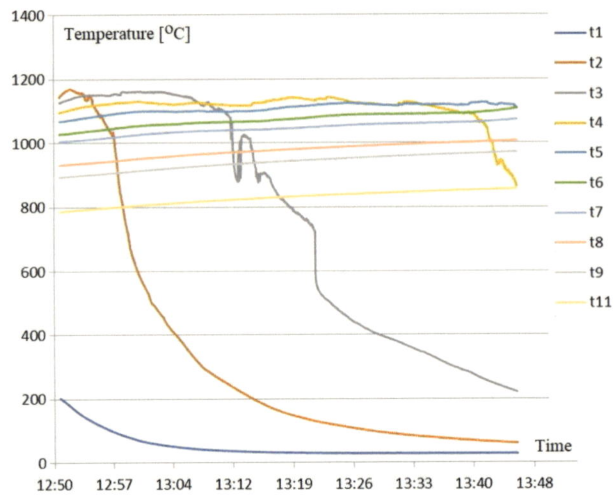

Figure 7.
Temperature values inside the combustion chamber, with $ZnCr_2O_4$ catalyst presence.

Overall, the started research is to be continued and further tests need to be done for simulating more different regimes, as well as to investigate the influence of catalysts at higher working temperatures.

2.2 Laboratory analysis on biomass substrates for determining their physical and chemical potential for different applications

This part of the present material will underline a part of the experimental determinations for different types of biomasses to determine their potential for anaerobic digestion or firing (co-firing) processes.

The used standards for laboratory determinations were:

- EN ISO 18134 – Solid biofuels. Determination of moisture content. Oven dry method;

- EN ISO 18122 – Solid biofuels. Determination of ash content;

- EN 14918 – Solid biofuels. Determination of calorific value;

- EN ISO 16948 – Solid biofuels. Determination of total content of carbon, hydrogen and nitrogen;

- EN ISO 16994 – Solid biofuels. Determination of total content of sulfur and chlorine;

- EN ISO 18123 – Solid biofuels. Determination of the content of volatile matter;

- CEN/TS 15370 – Determination of ash melting behavior.

The next table presents a small part of the determinations made for different types of biomasses (**Table 3**).

The chosen materials came from a very large selection, which stands as a base material for a database created by the first author of this chapter, database that is continuously under development and contains materials from agricultural, forestry,

DOI: http://dx.doi.org/10.5772/intechopen.101622

Excess air				[–]	2.0				
Burner power				kW	1.5				
Fuel ratio CH_4:CO_2				[–]	80.4:19.6				
Fuel lower heating value				[kJ/m^3]	28847.52				
Fuel consumption				[m^3/h]	0.1872				
Air consumption				[kg/h]	3.7041				
Burner pebble bed diameter				[mm]	13				
Burner pebble bed height				[mm]	253				
Burner pebble bed porosity				[–]	44.94				
Flame arrestor bed diameter				[mm]	6				
Flame arrestor bed height				[mm]	107.5				
Temperature distribution within the porous burner – gaseous phase									
−0.1	0	0.11	0.22	0.33	0.44	0.55	0.66	0.77	0.99
t_1	t_2	t_3	t_4	t_5	t_6	t_7	t_8	t_9	t_{11}
49.4	322	747	1108.7	1107.7	1073.2	1044.7	975.2	938.8	827.3
Temperature distribution within the porous burner – solid phase									
0		0.44		0.66		0.88			
t_{s1}		t_{s3}		t_{s4}		t_{s5}			
531.3		1065.2		1011.2		930.7			
Flue gas analysis at the exit of the burner									
O_2 [%]		CO_2 [%]		NOx [ppm]		CO [ppm]			
11.7		6.30		2		248			

Table 2.
Testing $ZnCr_2O_4$ catalyst.

Material	Observations	Source
Paulownia (Paulownia tomentosa)	pre-dried sawdust	entire plant
White poplar (Populus alba)	pre-dried sawdust	entire plant
Elephant grass (Pennisetum purpureum)	pre-dried sawdust	entire plant
Hemp (Cannabis sativa)	pre-dried sawdust	entire plant
Sunroot (Helianthus tuberosus)	pre-dried sawdust	entire plant

Table 3.
Types of analyzed biomass.

household, municipal, and industrial fields of application for all that involves biodegradable or partially degradable materials.

The next tables are going to present some general aspects concerning the properties of the studied materials and the potential application for energetic conversion (**Table 4**).

The moisture of the presented materials is considered for already pre-dried materials. From an ash content, the white poplar and sunroot have the largest values, making them not the first choice for firing processes, due to their high residue and ash content.

The calorific value is high for all the studied materials, and a very interesting aspect is the fact that the arborescent samples have net calorific values close to plant biomass, making them suitable for both energy conversion processes.

The carbon content and nitrogen are very specific for biomass, while the hydrogen content is close in value from one species to another, except hemp. There are no exceptional or different values than the ones expected for this type of material. Relative to C/N ratio, the best suitable biomass would be Sunroot, with a ratio of around 31. According to existing literature, the optimum domain for C/N ratio should be between 20 and 30, but from experience some materials do not meet these criteria and can be used for anaerobic digestion (**Table 5**).

The four specific points are very important to determine the specific temperatures at which the materials are starting to transform and reach a flowing point.

Material	Moisture content [%]	Ash content [%]	C [%]	H [%]	N [%]	Volatile content [%]	Gross calorific value [J/g]	Net calorific value [J/g]
Paulownia (Paulownia tomentosa)	10.04	1.12	45.6	6.44	0.329	84.16	20,218	18,659
White poplar (Populus alba)	10.7	5.93	43.6	6.4	0.92	76.8	19,350	17,764
Elephant grass (Pennisetum purpureum)	12.7	1.86	42.6	6.3	0.09	83.9	19,234	17,651
Hemp (Cannabis sativa)	15.5	2.87	48.3	5.4	0.45	79.1	19,334	17,940
Sunroot (Helianthus tuberosus)	11.3	5.94	45.3	5.59	1.46	74.3	18,277	16,903

Table 4.
Material energy properties, analysis on a dry basis.

Material	Shrinking temperature [°C]	Deformation temperature [°C]	Hemisphere temperature [°C]	Flow temperature [°C]
Paulownia (Paulownia tomentosa)	1080	1280	1370	1400
White poplar (Populus alba)	1010	1540	> 1540	> 1540
Elephant grass (Pennisetum purpureum)	730	1010	1250	1280
Hemp (Cannabis sativa)	960	1240	1300	1320
Sunroot (Helianthus tuberosus)	610	1280	1490	> 1540

Table 5.
Material chemical properties, analysis on dry basis.

In this regard, combined with the energetic values and the ash content, it can be determined a good behavior of materials from a firing process point of view.

The sunroot material presented an unexpected high value for flowing temperature, while the rest of the materials presented expected values. The bark in white poplar made that the specific flowing point to be of a high value, as estimated.

In the context of the presented materials, the bark is to be excluded from analyzed samples, and the high ash content is a parameter that determines if the materials are suitable for combustion, co-combustion or other processes.

Of course, there are other parameters to be considered, like chlorine, sulfur or heavy metals, when taking into consideration all the variables to the energetic conversion, but this is just a partial analysis and the conclusions are traced accordingly.

The main applications for the study was to determine the energetic potential of biomass types not usually applied for firing or anaerobe digestion processes and to study their potential application in those two directions. The materials were chosen because there is not enough literature to discuss different potential applications for Elephant grass, hemp or sunroot in anaerobic digestion or combustion, while Paulovnia and White poplar were chosen as comparative used materials, especially for firing applications. Present studies are made for anaerobic digestion of a part of the studied materials, but the work is still in progress.

2.3 Laboratory studies for biogas production and system development in terms of parameter monitoring and initial inputs for new different materials used for anaerobic digestion processes

This last part represents the focus of the research developed so far by the chapter authors. First, there will be depicted some of the small-scale test rigs developed so far, starting with commercial ideas, but less expensive and with good capability in terms of process control and results.

The next two figures present small-scale test rigs designed for preliminary testing of biogas production from different substrates (**Figure 8**).

The components found in the figure are:

1. Thermostatic bath with 6–8 places for heating the used materials for the anaerobic fermentation process (the temperature is controlled with the help of a thermostat and can be checked with the help of a thermometer inserted into the bath);

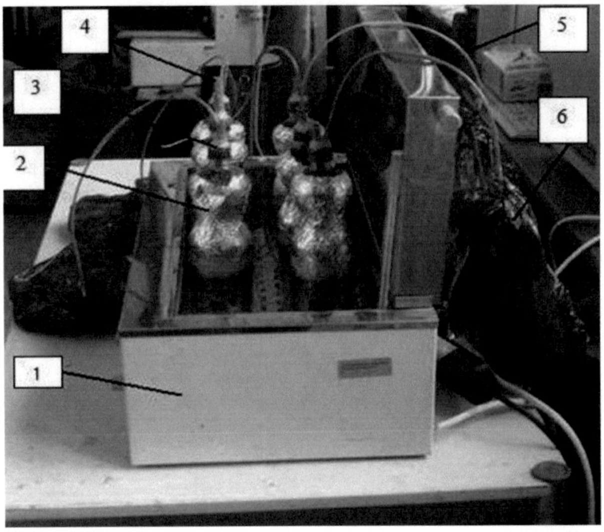

Figure 8.
Small scale test bench.

2. plastic bottles with a total volume of 1 (or 2) L, depending on the testing setup, filled up to about 0.8 (or 1.5) L with the materials used for determinations;

3. the corks of the plastic bottles were modified to allow both samplings for pH checking, homogenization using plastic syringes, and gas transfer from the bottles into the gas bags. Also, because of the light sensibility of the anaerobic bacteria, the bottles were covered with aluminum foil;

4. hose orifice for syringe insertion, used for sampling and homogenization;

5. connection (small diameter hose) between the plastic bottle and the gas bag for biogas storage;

6. gas bag for biogas storage.

The second small-scale test rig is dedicated to processes at ~4 L and allows better control for substrate agitation, while offering different levels for temperature (**Figures 9** and **10**).

Each part is a separate module composed of the reactor with lid, syringe for sampling and ph control and control panel for controlling temperature and agitation inside each reactor. In **Figure 11** there can be observed 4 modules that work independently from one to another. Both test rigs were used to determine biogas potential in terms of quantity and methane, carbon dioxide, hydrogen sulfide and dissolved oxygen for different recipes/substrates. The next part presents preliminary results for different experiments.

2.4 Laboratory production of biogas from waste waters, on 2 L scale test bench

The experiments were conducted in two batches. For the first batch, the used substrate materials were: waste water from urban treatment plant (M1), waste water from brew factory (M2), 95% waste water from treatment plant and 5% beet molasses (MM1), 95% waste water from brew factory and 5% beet molasses (MM2), 95% waste water from treatment plant and 5% cow whey (ZM1) and 95%

Figure 9.
Small scale modules.

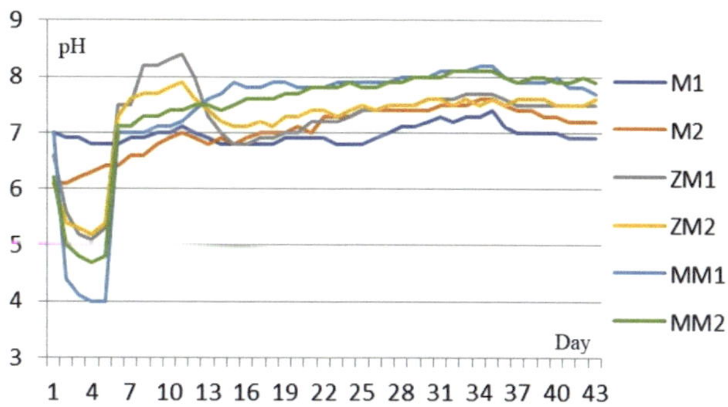

Figure 10.
Substrate pH variation in time, first batch.

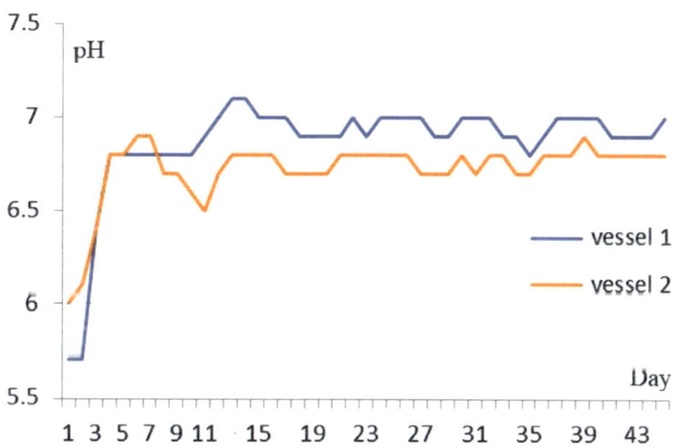

Figure 11.
Substrate pH variation in time, second batch.

waste water from brew factory and 5% cow whey (ZM2). The temperature regime was held at 36–37°C and the process parameters which were controlled consisted in pH, biogas partial composition and obtained quantities. The pH time variation for the material batches is presented in **Figure 12**.

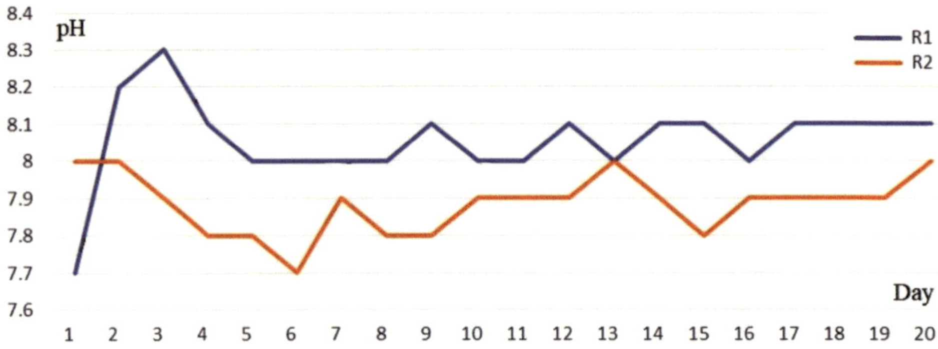

Figure 12.
Substrate pH variation in time, first batch on 4 L reactors.

From the pH variation, it can be observed that the co-fermented batches containing molasses for both residual sludges had the lowest pH, which increased after many corrections, which inhibited the fermentation process.

The pH for M1 and M2 was in the correct range and needed small interventions in terms of correction, the batch failed to produce a cumulative quantity of biogas that could be properly analyzed.

The most notable biogas quantities were produced by the MM1 (6 L) and ZM1 (4.5 L) batches. The maximum composition obtained after analyzing the produced biogas was for MM1, with 75% CH_4 and 10% CO_2.

On the second batch, the experiments were conducted in parallel, the second batch material used was formed as follows: 91% residual water from the treatment plant, 4% dehydrated sludge from the treatment plant and 5% cow whey for the first vessel and 91% residual water from the beer factory, 4% dehydrated sludge from treatment plant and 5% cow whey for the second glass vessel. The pH of the suspension was corrected with a solution of NH_3 20% concentration and the temperature regime was held inside the domain of 36–37°C.

It can be observed that during the process, the batches presented a relatively high pH value which made the use of the NH_3 suspension to be made just at the beginning of the process when the starting pH wasn't neutral.

Even if both batches produced biogas, the main composition of the produced gas until the end of the process was about 60 ÷ 61% CH_4 and 38 ÷ 40% CO_2 for both batches of material.

The produced quantities were about 4 L of gas for the mixture with residual water from the treatment plant, 4% dehydrated sludge from the treatment plant and 5% cow whey and about 5 L for the batch composed by residual water from the beer factory, and 4% dehydrated sludge from treatment plant and 5% cow whey.

2.5 Laboratory production of biogas from waste waters, on 4 L scale reactors

These experiments were also conducted in two batches. For the first batch, two reactors were used which had a total volume of 5 L each, of which 4 L was the useful volume. The temperature of reactor 1 (TR1) was 37°C and the temperature of reactor 2 (TR2) was 42°C. Reactor 1 (R1) had a 3.5 L suspension consisting of a specific mixture of biogas with corn silage and wet fraction plus 100 g of degraded maize grains. Reactor 2 (R2) had a 3.5 L suspension consisting of a specific mixture of biogas with corn silage and wet fraction plus 100 g of potato peel.

The figure above shows the pH levels for the two reactors used within 20 days of the experiments. It can be observed that in the first phase (the first 7 days the

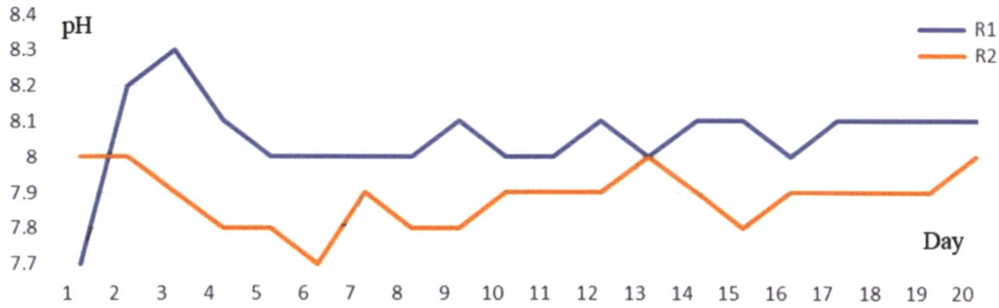

Figure 13.
Substrate pH variation in time, second batch on 4 L reactors.

pH has a slightly decreasing tendency varying between 7.8 and 8.3, in this sense, it is observed that the substrate used is very stable over time even in the initial phase which has as specific a relatively acidic pH (below 6).

Throughout the process, the pH remains slightly alkaline for both reactors, being a good indicator for optimal development of the anaerobic fermentation process.

The first reactor was heated in mesophilic mode, and reactor number 2 in thermophilic mode (in its lower range), and in correlation with the pH identified during the experiment the elements of influence are favorable to produce a quantity optimal biogas with a relatively high methane concentration.

The cumulative amounts of biogas for the two reactors identify a value of about 14 L of biogas for reactor number 1 and about 8 L of biogas for reactor number 2. The difference in quantity between the two reactors can be explained by the temperature regime applied to each reactor in part (the thermophilic regime has as specific a higher production in the time of biogas) and the slightly different composition because the potato peel has a high starch content, an aspect that can be beneficial but also inhibitory, by the appearance of the foaming phenomenon, depending on the behavior of each load separately.

On the second batch, two reactors were used which had a total volume of 5 L each, of which 3.5 L was the useful volume. The temperature of reactor 1 (TR1) was 37°C and the temperature of reactor 2 (TR2) was 42°C. Reactor 1 (R1) had a 3.5 L suspension consisting of a specific mixture of biogas with maize silage and wet fraction of animal biomass plus 100 g of degraded maize grains. Reactor 2 (R2) had a 3.5 L suspension consisting of a specific mixture of biogas with corn silage and the wet fraction of animal biomass plus 100 g of potato peel (**Figure 13**).

For batch number 2, a similar pH behavior is observed for the two reactors with maximum values of about 8.3 for R1 and 7.9 for R2. Under certain conditions, pH values above 8 can slightly inhibit the biogas production process, but some substrates react positively to slightly higher pH values (between 7.5 and 8).

The temperature regime chosen is similar to that of the first batch, again noting that the heating system ensures a relatively constant temperature throughout the process for both reactors, with a difference of up to one degree compared to the desired operating temperature. Reactor number 1 produced about 14 L of biogas while reactor number 2 produced about 8 L of biogas, in this case, both the higher pH values and the slightly higher temperature range high being a negative influencing factor on the anaerobic fermentation process.

The concentration of methane for reactor number 1 is about 49% while for reactor number 2 it is about 51%, values which again are an indicator of a low potential for the use of independent in combustion processes.

3. Conclusion

Relative to the presented material, the further conclusions can be traced:

- The most important catalyst with positive results in terms of the combustion process is $ZnCr_2O_4$ and further determinations must be made to determine if other catalysts can be used with better results than the existing ones.

- The analyzed materials have good energetic potential which can be further studied in terms of firing or co-combustion processes and anaerobe digestion for biogas production – chemical analysis is the most important one to determine further process influence in this regard.

- Small-scale bioreactors were developed in-house after ideas collected from literature and existing experience in practical determinations at the laboratory level.

- During testing, temperature and pH proved to be very important to maintain a live microbiota and increase the anaerobic digestion process speed in time, with good indicators relative to biogas quantity and volume participation of methane (over 50% allows firing process).

All the presented elements, even if presented separately, are interconnected to add plus value to a known conversion process to further bring new perspectives in terms of used substrates, increased quality for the produced biogas in terms of methane concentration and further applications in firing processes to maximize the energetic output conversion to heat or electric energy.

Acknowledgements

The authors of this material wish to recognize the support of the following entities and colleagues to the presented results. The catalytic firing process was made together with collaborators from Mechanical Engineering Faculty in Belgrade, and the authors want to address special acknowledgement to Prof. Srba Genic and Ass. Prof. Mirjana Stamenic for all the support in this regard. The laboratory determinations for biomass were conducted with the team from BEA Institute for Bioenergy, Vienna, Austria.

Conflict of interest

The authors declare no conflict of interest.

DOI: http://dx.doi.org/10.5772/intechopen.101622

Author details

Adrian Eugen Cioabla and Francisc Popescu*
University Politehnica Timisoara, Timisoara, Romania

*Address all correspondence to: francisc.popescu@upt.ro

References

[1] EEA. Annual European Union Greenhouse Gas Inventory 1990-2019 and Inventory Report 2021 [Internet] 2021. Available from: https://www.eea.europa.eu/ds_resolveuid/f454ae4c825646b2a15497b17a71dbf0 [Accessed: October 10, 2021]

[2] Covaliov V. Improvement of biogas production systems. Akademos. 2015;**4**(39):38-44

[3] Kerroum D, Mossaab B-LH, Hassen MA. Production of biogas from sludge waste and organic fraction of municipal solid waste. In: Kumar S, editor. Biogas. Rijeka: Intech; 2012. pp. 152-172. DOI: 10.5772/32718

[4] Nandi R, Saha CK, Sarker S, Huda S, Alam M. Optimization of reactor temperature for continuous anaerobic digestion of cow manure: Bangladesh perspective. Sustainability. 2020;**12**(21): 8772. DOI: 10.3390/su12218772

[5] Czubaszek R, Wysocka-Czubaszek A, Banaszuk P. GHG emissions and efficiency of energy generation through anaerobic fermentation of wetland biomass. Energies. 2020;**12**(24):6497. DOI: 10.3390/en13246497